PHYSIOLOGICAL ASPECTS OF WATER AND PLANT LIFE

by

W. M. M. BARON, B.Sc., M.A.

Winchester College

HEINEMANN EDUCATIONAL
BOOKS LTD: LONDON

Heinemann Educational Books Ltd
LONDON EDINBURGH MELBOURNE TORONTO
SINGAPORE AUCKLAND JOHANNESBURG IBADAN
HONG KONG NAIROBI NEW DELHI

ISBN 0 435 61054 6

First published 1967
Reprinted 1971

Published by
Heinemann Educational Books Ltd
48 Charles Street, London W1X 8AH

Printed in Great Britain at
the Pitman Press, Bath

Contents

List of Plates

Preface

The subject of plant water relations has never been a particularly easy or popular one. Perhaps this may be because many of the phenomena involved seem to be, at least at first sight, obvious and commonplace. Yet closer investigation quickly shows how wrong this can be, and a thorough understanding of the physical and chemical properties of the water molecule is necessary if we are to understand many of these processes. Maximov's book *The Plant in Relation to Water* published in 1929 was the first to examine the various problems in detail and I have still found it useful to draw extensively from his ideas. In recent years research has once again revolved around this topic and two symposia have helped to revitalise interest. The work of Professor Weatherley at Aberdeen has been of particular value in dealing with the important topics of the osmotic relations of the cell and the movement of water in the plant. I am most indebted to the help he has given me and for allowing me to quote from his ideas and results.

In a book of this size it has not been possible to include a number of aspects of water relations that are important. For instance, relatively little space is given to plant anatomy; likewise the topics of water in the soil and mineral uptake receive less mention than I would have liked. Excellent references to these topics are quoted in the bibliography.

I would like to express my thanks to my wife for her help, advice and for providing Plate 5. I am also indebted to the Alpine Garden Society for providing Plate 1, to Dr. B. E. Juniper for Plates 3a and 3b, to *Amateur Gardening* for Plate 6, to Mr. W. H. Dowdeswell for Plate 7 and to the United States Information Service for Plate 8. A full acknowledgement and reference list for the various figures and tables are given in the bibliography (p. 137) and the acknowledgements (p. 139).

My thanks are also due to Mr. W. H. Dowdeswell, the editor of the series, for providing encouragement and assistance at all stages and to Mr. G. V. Darrah for reading the typescript. Lastly I must thank Mr. H. MacGibbon of the publishers whose enthusiasm and patience have been of the greatest help.

Winchester, December 1966. W.M.M.B.

1

The Properties of Water

Life and the properties of water

It is only in times of water shortage that man begins to realise how totally dependent all living things are on water. Because it is the only common liquid on our planet, we are almost over familiar with it and tend to forget many of its remarkable properties, on most of which our very existence depends. Indeed, rainfall is probably the most important single environmental factor determining plant growth and animal survival.

Life, at least as we know it, can only exist in a narrow range of conditions and these are determined to a large extent by the properties of water. Water plays a great part in the maintenance of environmental stability. The thermal properties are particularly important here; water has a high specific heat which gives it a marked capacity for heat storage. The oceans, lakes, rivers, and water vapour in the atmosphere all help to offset severe temperature fluctuations.

The high density of water above freezing point is another unusual feature and means that life can continue under water when the surface is frozen. This is more significant now than it was when life was confined to the warmer seas in the early days of evolution. The expansion of water when it freezes has one detrimental effect. Most plant and animal cells are composed principally of water and so are liable to be badly pamaged if ice crystals form in them.

Water has other important characteristics. It is a unique solvent and provides a medium for reactions between ionised as well as unionised substances and is therefore often regarded as a catalyst of chemical reactions concerned with cell metabolism. It can allow for the stabilisation of large molecules such as proteins; it is an excellent medium for providing support and transport and its strong cohesive properties allow for the maintenance of water columns through the plant. To understand some of these characteristics it is necessary to know the details of the structure of the water molecule itself.

The structure of the water molecule

Hydrogen and oxygen on their own are rather reactive substances. Even water itself is a much more complex substance

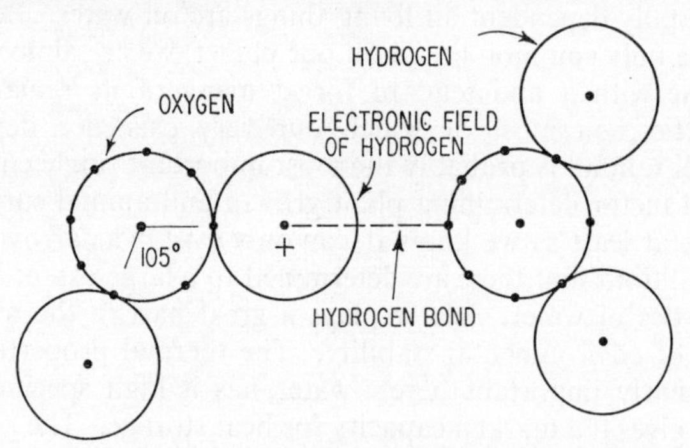

FIG. 1.1. Water molecules and the hydrogen bond

than might be expected and can exist in a wide range of forms and enter into a variety of reactions. Fig. 1.1 represents two molecules of water and serves to illustrate some salient points about its structure and properties. First, each water molecule is polar in nature, for the hydrogen atoms reach a stable position in relation to the oxygen, being on one side of the molecule at an angle of 105° to each other. This polarity enables hydrogen bonds to form between any hydrogen nucleus, with its positively charged proton, and another electron. The bond forms on the unoccupied side, as it were, of the oxygen atom. In the diagram the hydrogen bond is represented by a dotted line between the two water molecules. This arrangement provides a satisfactory explanation for features of cohesion and surface tension which are so crucial to plant and animal life.

Apart from the different forms of water based on the different isotopes of hydrogen (H^1, H^2, H^3) or of oxygen (O^{16}, O^{17}, O^{18}), water exists in a number of forms with different molecular and ionic arrangements. In ice and snow, adjacent water molecules are packed in an orderly array, linked symmetrically by hydrogen bonds. In liquid water the arrangements are much more variable; clusters of molecules may share hydrogen bonds which must be broken if the liquid is to flow. Finally, water may exist as true ions—mainly hydrogen (H^+) and hydroxyl (OH^-) but also oxygen (O^{--}). Although these ions have some degree of cohesion they are still able to vibrate freely in response to changes in temperature—a property which accounts for the high specific heat of liquid water.

Ions, hydrides and hydrates

Unionised substances may or may not dissolve freely in water. Water has the highest *dialectric constant* known. This means that it has a high capacity for neutralising charged particles and so allowing ionised substances to dissolve. If common salt is dissolved in water the ions are effectively kept apart by shells of water molecules which surround them; sodium and chloride ions *polarise* the water molecules (see Fig. 1.2). The vacuole and cytoplasm of living cells therefore contain a wide range of dissolved substances that may be useful in metabolism.

Proteins, which are a vital part of the content of plant cells, may depend on water for their structural organisation through the formation of hydrogen bonds across parts of their molecules. If the water is removed the protein will be denatured, though the effect is not necessarily permanent and return to normal hydration often results in the protein regaining its proper organisation. In this respect water serves a similar function to the plasticisers used in plastics. *Hydrides* such as

HCN and NH_3 owe their poisonous properties to the fact that they form stable associations with proteins which render them permanently denatured.

Recently some electrically neutral substances of the right size and shape have been found to collect water molecules around them in an orderly way. These associations are called *hydrates*. Proteins contain groups which can form hydrates; these may have disastrous or useful effects on the cell. Hydrates sometimes have a lower density than ice. So their formation may cause a destructive expansion of the cells that gives

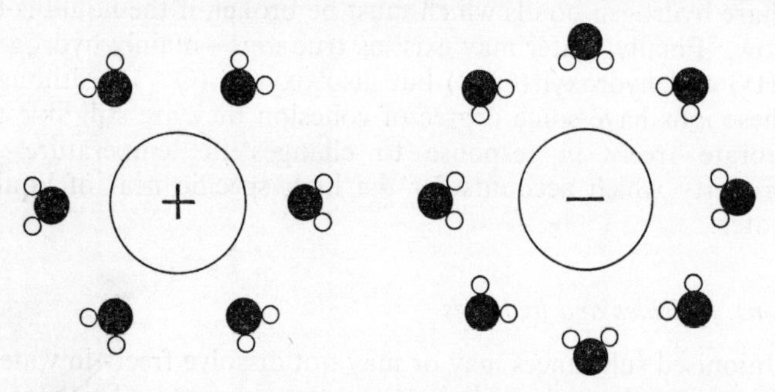

FIG. 1.2. Ions and their surroundings. Ions attract water molecules to them; this polarises the water molecules and so keeps the ions apart

the appearance of frost damage. Such effects have been noticed in maize (*Zea mais*) at a temperature a few degrees above freezing. On the other hand *slow* deposition of hydrates need not result in cell destruction and can indeed prevent freezing of cell materials should low temperatures occur, providing a sort of built-in antifreeze compound for the plant.

Colloidal systems

In addition to the substances in true solution, whose particles are defined as possessing diameters of no more than 1 mμ (1/1000th micron), water also provides the dispersion medium for larger particles. If these particles are larger than about one micron the system would be regarded as a *suspension*, but if smaller than this it would show *colloidal properties*. Colloids exhibit a number of important properties and owe their stability to several factors. Colloidal particles are small enough to be affected by the thermal activity of the ions and molecules in the dispersion medium. In addition their particles may be electrically charged and repel one another. The activity of such particles can be seen under the microscope and is called *Brownian movement* after Robert Brown who first observed it, in 1826, while studying the spores of the clubmoss *Lycopodium*.

In most cytoplasmic systems the dispersed colloidal particles are often hydrated by a shell of water molecules, like the ions already discussed. These constitute the *lyophilic* colloids, which have a considerable stability, brought about by the electrical attraction between the particles and the water molecules.

Colloids may exist as either *sols* or *gels*. These differ largely in that the latter are much more viscous. This occurs through the molecules of the dispersed particles being large and perhaps forming some sort of network. Alternatively, the quantity of the dispersion medium may be reduced. The imbibing powers or *imbibition potential* of a dry or partly dehydrated gel is extremely strong and accounts for the ability of dry seeds, seaweed fronds and some other tissues to absorb great quantities of water at certain metabolic stages. While the cytoplasm in a normally active cell is mainly in the state of a sol, where the fluid condition of the dispersal medium predominates, some visible parts such as the starch grains are true gels.

The small size of colloidal particles gives them a relatively large surface area and consequently a high surface energy; this may increase the chemical reactivity of the system of which they are part. When two immiscible liquids (or solid and liquid or solid and gas) are in contact *adsorption* of dissolved particles takes place at the adjoining surfaces. In this way colloids can adsorb extremely large concentrations of various materials.

Water and the cell

The foregoing sections show that the cytoplasm must be fully hydrated if it is to be metabolically active. In a fully hydrated state the various dissolved substances diffuse freely

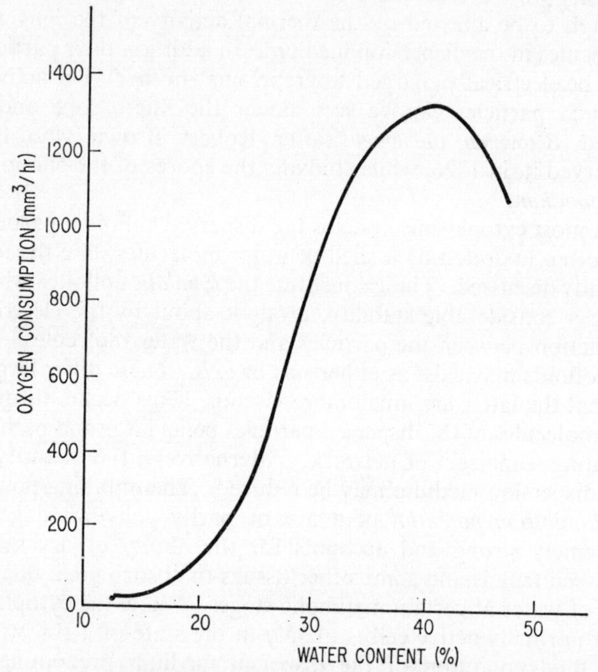

FIG. 1.3. The effect of water content on the rate of respiration of germinating seeds of oat (*Avena sativa*)

from one part of the cell to another and so take part in the various reactions that are going on. Again, the larger colloidal particles must be fully hydrated if they are to exhibit their adsorptive capacities to the full. The work of Bakke and Noecker (1933) on germinating seeds of the oat (*Avena sativa*) has shown how hydration and metabolic activity are related. Fig. 1.3 shows that a water content of about 40 per cent provides an optimum rate of respiration for oat. With a water content of only 20 per cent, respiration, providing the energy needed for the rapid growth and germination of the seed, is barely under way. Water enters into a number of other well-known reactions in the cell. The primary stage of photosynthesis can be taken to be the photolysis of water. The reducing potential released by this photolysis can be used in the reduction of carbon dioxide to form a carbohydrate. In respiration, on the other hand, the vital energy releasing reactions are those which involve the passage of hydrogen ions through redox chains so that oxygen can be reduced and water formed. Metabolism also involves numerous condensing and hydrolysing reactions, such as the formation and breakdown of starch.

The relationship of the cytoplasm to the vacuole is another important aspect of cell organisation. The vacuole develops quite early on in cellular differentiation and presumably involves the laying down of an inner plasma membrane or *tonoplast* which separates the two liquids. Unlike the cytoplasm the vacuole contains relatively few visible and colloidal particles but many dissolved substances. In many cases cell expansion is in great part due to enlargement of the vacuole, so the cytoplasm and its membranes must have the capacity to allow water to accumulate as well as dissolved substances. How this occurs is discussed in Chapter 3 (p. 18) on water uptake. Contractile vacuoles which occur in a number of the more primitive algae prevent excessive water uptake and are probably under direct metabolic control.

Clearly then the importance of water to plant life is immense. Water is of fundamental importance for the cell to carry out the functions of metabolism. In addition, attention has been focused on the overall water turnover of the plant and its tissues; the means by which water is absorbed, lost and serves to transport substances. Finally, it provides a remarkably stable environment without which it is doubtful if life as we know it could exist.

2
The Uptake of Water

The magnitude of water uptake

Much of the impetus for research into water uptake in the last century derived from the hope of increasing the performance of various crops growing in water deficient soils. This aim has only partially been realised. We now have a fair understanding of the process of water uptake, but more ecological work is still to be done on plants growing close to one another under natural and cultivated conditions.

The factors that influence water uptake are largely those that determine water loss from the leaves. Sachs in the 1870's investigated the quantity of water absorbed by the roots and shoots. He estimated the water absorbed by tobacco plants over five days, the roots and shoots being separated and investigated individually. The roots exuded water and yielded 16 ml during the whole period while the shoots absorbed 200 ml. Root exudation is a poorly understood phenomenon that is known to be at least partly under metabolic control. Water uptake depends on the size and shape of the plant as well as the prevailing environmental conditions. In some plants absorption may be only a few ml per day, in others, such as large trees, it may be several hundred litres.

More critical measurements of the quantity of water absorbed by the plant were carried out by Vesque in 1876. He devised a number of potometers, one of which is shown in Fig. 2.1, to measure water uptake by timing the flow of the meniscus along the capillary tube. Water loss or transpiration occurs in more or less similar amounts to water uptake. There are a number of difficulties in designing and setting up the apparatus; the reservoir must be small or temperature fluctuations will affect the results. Rooted plants, obviously the most desirable, can be obtained by rooting cuttings in water or by digging up rooted plants from the ground, but neither of these

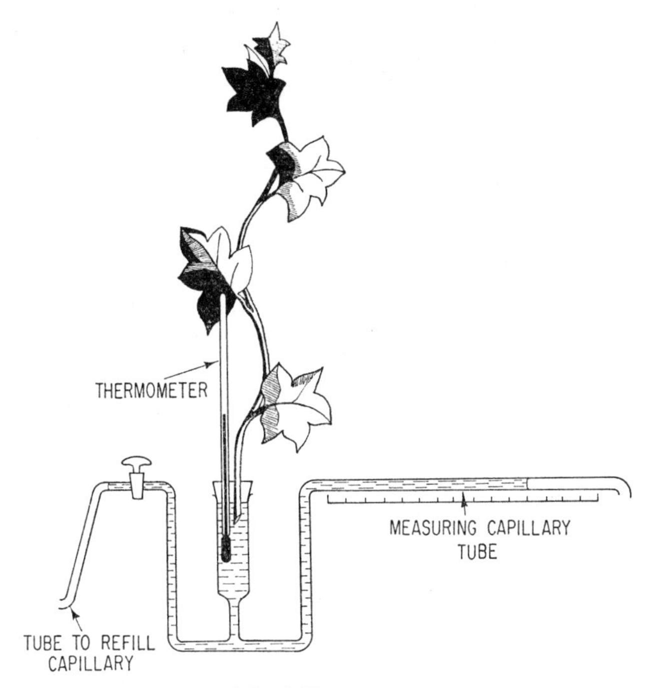

FIG. 2.1. A Vesque potometer

techniques preserves the plant in its natural state. This is even more true of shoots.

Other designs of potometer measure both water uptake and water loss simultaneously. This permits us to distinguish water used for *internal* physical and metabolic purposes from those concerned with the replacement of water lost from the aerial parts by transpiration. A simple method often used is to weigh the whole plant with its roots in a tube of water. Water loss will be given by loss in weight. If the side of the tube is graduated, uptake can also be measured. In practice the values obtained usually come close together. Fig. 4.5 shows the sort of results obtained from such an experiment.

The root system and the site of water uptake

Any natural community which is fully established has its various tree, shrub and herb species clearly zoned or stratified. A number of factors may determine these zonations but one of the least well understood is the relation between zonation of roots in the soil and the distribution of species. Fig. 2.2 shows how roots are zoned in a chalk heathland soil. Here pH as well as water content varies along the section. There is a complex distribution of the roots of different species, but they mix only to a slight degree. Fig. 9.5 shows a slope covered with Mediterranean vegetation, where the niches occupied by the roots of different species are again distinct.

It can be seen from these figures that root profiles vary in appearance from species to species. In Fig. 2.2 the salad burnet (*Poterium sanguisorba*) has a little-branched tap root, while the ribwort plantain (*Plantago lanceolata*) has a branched tap root and the grass has a fibrous system. The root of the ling (*Calluna vulgaris*) is mycorrhizal and highly branched.

If a root measuring more than 3–4 mm in diameter is stained with Sudan it becomes apparent that a great deal of cutin and suberin is deposited in the outer layers (epidermis). These substances are relatively impermeable to water and so it seems likely that water absorption must take place nearer the apex. In many plants the zone of root hairs begins 2 or 3 mm behind the apex of the root and lasts about another 15 mm. The root apex itself is uncutinised but some cutinisation often begins in the older parts of the root hair zone. Water absorption takes place through both root hairs and the relatively unspecialised cells of the root apex.

Uptake through other parts of the plant

Land plants take up most of their water through the root system but a number of lower plants—littoral and aquatic

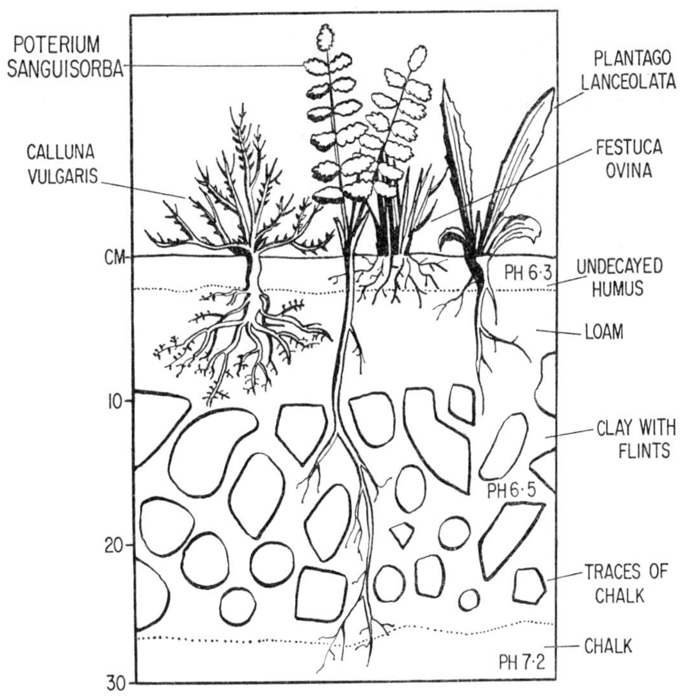

FIG. 2.2. Zonation of roots in a chalk heathland soil

algae, fungi, mosses and liverworts—can take up water through almost any part of their surface. Relatively little is taken up through the rhizoids of the smaller mosses and liverworts, the principal absorptive structure here being the leaves or thallus, though in the larger species of moss, considerable absorption takes place through the rhizoids. Aquatic flowering plants such as the water crowfoot (*Ranunculus aquatalis*) also absorb water through their leaves. Absorption of water through fern leaves may also be considerable, though becomes limited by cutinisation. A classic case here is the resurrection plant (*Selaginella lepidophylla*) which revives from a curled-up ball of leaves in a few minutes after these are wetted. This plant lives in central America, from Texas to Peru where ability to regain metabolic activity during the seasonal rains may be crucial for survival. Recent methods of plant propagation involving mist spraying of leaf areas have shed much light on the absorption of water by conifer and flowering plant leaves with cuticles which are apparently impermeable. It would seem that the stomata are not the only possible site for foliar absorption. In some Conifer species, e.g. the Scots pine (*Pinus sylvestris*) the leaves arise from a distinct scaly sheath. Recent work by Leyton and Juniper (1963) has shown that this sheath area may be a water absorbing area.

TABLE 2.1

Uptake of water by pine needles in October

(Percentage of initial weight after 24 hours; means of six measurements ±S.E.)

Depth of immersion of pine needle in water (cm)	*Uptake*	
3	−0·16	±0·57
6	+0·61	±0·48
9 (1·5 cm below sheath)	+1·38	±0·42
12 (1·5 cm above sheath)	+6·53	±1·12
Through cut bases of needle	+7·30	±0·35
Through cut tips of needle	+9·52	±0·47

Many epiphytes absorb the greater part of their water through their stems and leaves. A famous example is the Spanish moss (*Tillandsia usneoides*), a curious flowering plant that is common in tropical and semi-tropical America. Its long, straggling stems and leaves are covered with shield-like hairs that allow the penetration of water but, when dry, shrink and prevent water loss.

Where plants like the resurrection plant are living under conditions of drought, water uptake from dew may be important for plant survival. In a recent survey on the importance of dew, Monteith (1963), it is shown that semi-xerophytes of Mediterranean countries, like Aleppo pine (*Pinus halepensis*) and olive (*Olea europaea*), can take up considerable quantities of water from dew to counteract the saturation deficit of their leaves. This daily moisture supply, though slight, is utilised directly at the site where the dew forms, and is an essential part of the water economy of such xerophytes.

TABLE 2.2

Uptake of water from soil and atmosphere by Aleppo pine and olive

Saturation deficit = 100 × (Saturated leaf weight − Actual leaf weight)/Saturated leaf weight

Time	*Saturation deficit* (%)
Evening	19
Following morning:	
Dewed	10
Undewed	17

The soil solution and some special situations

Water uptake must first satisfy the needs of the individual cell or the plant will wilt through loss of turgor and eventually die. This uptake will be partly osmotic and will therefore be affected by the osmotic potential of the soil solution. Drier soils will have a higher concentration of solutes than wet ones and as they become drier uptake will become progressively more difficult. Shull in 1916 investigated this problem using seeds of the cocklebur (*Xanthium pennsylvanicum*) which swell rapidly in moist conditions. Table 2.3 summarises his results.

The water retaining capacity of soil that will cause wilting in most mesophytes is of the order of 3 or 4 atmospheres, much less, in fact, than the figures in the table below would seem to suggest. However, many plants seem to be able to increase their water diffusion potential as dry conditions approach and

TABLE 2.3

The relation between water content of the soil, water uptake by seeds and the osmotic potential of the soil solution Xanthium

Water content of the soil (% absolute dry weight)	*Uptake by seeds (% air-dry weight)*	*Osmotic potential of the soil solution (Atmospheres)*
5·8 (air-dry)	0·0	965
9·4	6·5	375
11·8	11·9	130
13·2	21·4	72
17·1	37·7	19
18·9	47·3	3·8

the water potential (see p. 19) of a given plant may determine its ability to compete with another whose roots are growing in close proximity.

Briggs and Shantz in 1911 carried out an extensive series of investigations to determine the relative efficiencies of different plants in respect of their water economy. The plants were kept in a water saturated atmosphere in glass pots that were carefully sealed with wax. The investigators introduced the concept of *wilting coefficient of the soil.* This is the water content of the soil expressed as a percentage of its dry weight at the point when *permanent wilting* first takes place. Permanent wilting is when the leaves are unable to recover, even though the atmosphere is saturated, unless the soil moisture content is raised. The table below lists their results. The wilting coefficients are expressed in relation to the average value for the whole series. The average value is corrected to unity. Thus values in excess of unity indicate a relatively inefficient plant, while those less than unity show that the plant has a more efficient water economy.

TABLE 2.4

The relative wilting coefficients of different plants

Plants	*Relative wilting coefficient (average)*	*Number of observations*	*Probable error of average value*
Wheat (*Triticum*)	0·994	653	±0·002
Barley (*Hordeum*)	0·97	60	±0·006
Rice (*Oryza*)	0·94	21	±0·012
Various Papilionaceae	1·01	138	±0·005
Tomato (*Lycopersicum*)	1·06	20	±0·009
Taro or coco (*Colocasia*)	1·13	19	±0·005

Most plants show a more or less comparable efficiency, but those plants with coarse, slightly branched roots, such as members of the *Papilionaceae* and the tomato, seem less efficient than those with the finely branched fibrous roots that are usually found in the grass family.

Pioneer studies in the supply of soil water to the plant were made by Livingston in 1908. His auto-irrigator is illustrated in Fig. 2.3. The amount of water taken up by the soil is read from the change in level of water in the supply vessel. The water taken up by the plant, which is identified with the water loss, is obtained by weighing the whole system.

If a plant is taking up a great deal of water it would seem likely that a depletion of soil water may occur around the roots. Livingston found that this was so and that the soil may take some time to regain equilibrium after the plant's maximum period of uptake has passed. A typical situation is illustrated diagrammatically in the graph Fig. 2.4.

Plants growing in soils with abnormally high concentrations of soil solutes are faced with great problems of water supply.

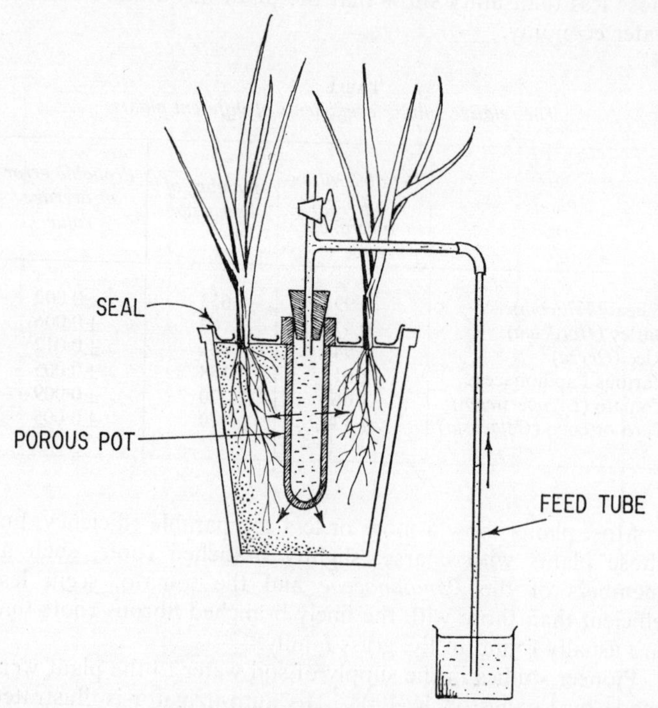

FIG. 2.3. Livingston's auto-irrigator

These high concentrations may be due to man-made factors such as faulty irrigation where climatic conditions are otherwise normal. In the Indus valley, years of badly controlled irrigation have resulted in mineral salt crystals appearing on the soil surface. The water table is too near the soil surface and evaporation has been rapid, so that the osmotic potential of the soil solution has progressively increased. There are recorded cases of this so called 'bitter earth' simply having to be bulldozed away once this has occurred. Similar situations

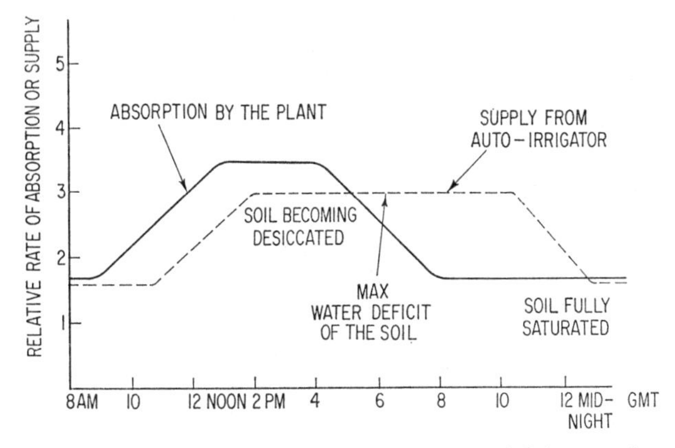

FIG. 2.4. Development of a temporary water deficit around a rapidly transpiring plant (diagrammatic)

occur in natural salt marshes. Plants capable of survival in these circumstances need to have high osmotic potentials as well as other adaptations which may enable them to survive in their rather inhospitable environment.

3

The Mechanism of Water Uptake

Water absorption by the cell

The uptake of water into a typical plant cell is in part due to the physical process of osmosis. By osmosis, water (or a similar solvent) passes through a semi-permeable membrane to dilute a stronger solution on the other side of the membrane. In a cell the vacuole contains dissolved organic substances and inorganic ions whilst the cytoplasm serves as the semi-permeable membrane. The cell wall itself is fully permeable, offering no barrier to the passage of water.

The osmotic properties of these parts of the cell can be studied by placing the tissue in a strong solution of potassium nitrate, which causes *plasmolysis*. Water passes from the dilute solution of the vacuole, through the cytoplasm and out into the stronger external solution. This results in a decrease in size of the vacuole. The cytoplasm comes away from the cell wall but continues to line the vacuole. In spite of its elasticity the cell wall only shrinks slightly and the external solution is seen to penetrate inside the wall. This is clear evidence that the cytoplasm is the semi-permeable membrane and not the cell wall, which seems to be fully permeable to both water and dissolved ions.

Fig. 3.1 shows plasmolysis in the cells of the staminal hairs of the spiderwort (*Tradescantia virginiana*). This species is a useful subject since the vacuole is coloured red by water-soluble purple anthocyanins. As plasmolysis takes place and solvent is removed from the vacuole, the colour of the anthocyanins becomes more intense. Red-coloured onion bulb scale cells are also useful demonstration material. Prolonged plasmolysis may permanently impair the workings of the cell so that it dies, but over short periods it is a reversible process and re-immersion of the tissue in water or dilute solution after a short time results in the cells regaining their normal organisation.

The pressures acting upon water present, entering or leaving the cell have been clearly divided by Weatherley (1952) into

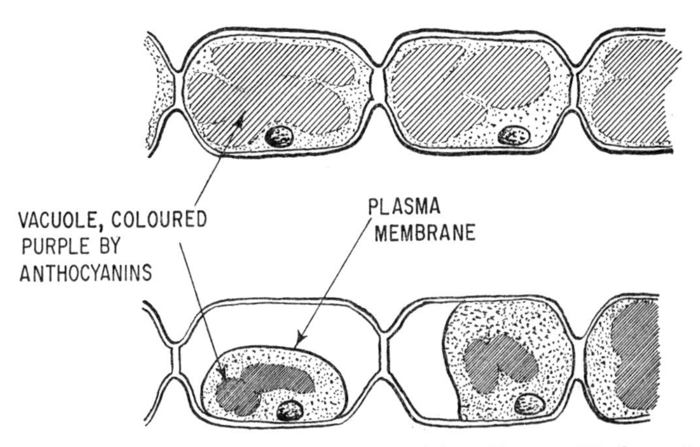

FIG. 3.1. Plasmolysis in staminal hair cells of the spiderwort (*Tradescantia virginiana*). Top: Normal cells. Bottom: Plasmolysed cells

hydrostatic pressures and pressures due to the presence of solutes.

The movement of water or *water diffusion potential* (often abbreviated to *water potential*) into the plant cell depends on two groups of factors; first the *osmotic potential* due to the presence of solutes in the cell sap. Second, the true hydrostatic pressure exerted by the elastic cell wall on the cell contents (*the wall pressure*). The resultant of these two forces gives the water diffusion potential of the cell. The water diffusion potential is the same as the older term suction pressure. The term diffusion may be in itself misleading for it is uncertain that osmosis is an entirely diffusional process and so the term used is now simply *water potential.* Fig. 3.2 summarises these relationships in a hypothetical plant cell or tissue. The wall pressure and the osmotic potential are plotted separately from their resultant—the water diffusion potential.

The curve ABC represents the gradual decrease in the

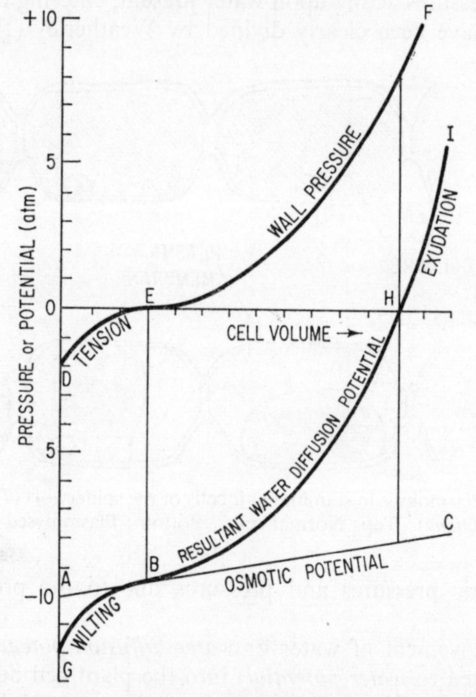

FIG. 3.2. Water potential in a hypothetical cell or tissue under various conditions

osmotic potential of the cell sap as the cell takes up water and the dissolved substances in the vacuole become diluted. The curve DEF shows the change in tension (negative pressure) or positive pressure exerted by the cell wall as the cell takes up more water.

GBHI is the resultant of these two factors. At B where plasmolysis is just beginning (*incipient plasmolysis*) the wall pressure is nil and the osmotic potential must be the same as the water potential. At H the positive hydrostatic pressure due to the cell wall exactly balances the negative osmotic potential. Thus there is no further uptake and the cell is fully turgid. Between incipient plasmolysis and full turgor the force distending the cell wall, equal and opposite to the wall pressure is referred to as the *turgor pressure*. In some instances the wall pressure may be increased after full turgor is reached. This results in exudation from the cells (see Chapter 4, p. 42).

Below incipient plasmolysis the cell wall has no effect on the water potential of the cell, but if the cell loses water by evaporation this may set up a negative pressure which results in a tension in the cell wall and consequent wilting and collapse.

We see then that there are two factors affecting the water potential of the cell; the wall pressure and the osmotic potential. The resultant between these two, the water potential, gives a measure of the capacity of the cell to take up water.

Considerable water uptake may also take place *not directly into the living cells*. Even with a fully turgid cell system water may pass in from the soil to make up for a negative hydrostatic pressure at the fine xylem endings caused by transpiration. The intercellular spaces as well as wood tissue provide the channels for this type of water uptake. The tightly packed root apex, however, has few intercellular spaces. Here the passage of water may occur through the only fully permeable part of the cell, the cell wall. Such areas are referred to as *apoplast* or *free space* as no metabolic processes are involved in mass flow through them. This situation may not be so far away from the long-since discarded ideas of Cesalpino, who interpreted water uptake in plants exclusively in terms of capillary attraction.

Differences in the water potential of the root cells of different species of plants may have considerable ecological significance. Movement of water within the plant may also be affected by the distribution of differing water potentials of various cells and tissues. Consequently the measuring of these potentials are of great importance both to plant ecologists and physiologists.

Measurement of osmotic and water potential at incipient plasmolysis

Reference to Fig. 3.2 shows that there is only one point at which the osmotic potential of the cell sap can be measured easily. This is at incipient plasmolysis. Here the osmotic potential will be the same as the water potential. Thin slices, epidermal strips, or whole leaves of thin-leaved plants (e.g. *Elodea* and mosses) are placed in sucrose solutions of differing strengths and covered to prevent evaporation. Those cells, in which the external solution is more dilute than the cell sap, take up water, those in which the external strength is greater lose water and become plasmolysed. If about half the cells are just showing plasmolysis, then the tissue is regarded as at incipient plasmolysis and the strength of the external solution is said to be *isotonic* with the cell sap. In practice it is difficult to estimate this point accurately. The strength of dissolved materials in the various cells of the same tissue may vary and some may have been damaged in making the preparation. Careful counts of a large number of cells must always be made, and the tissue must not be left too long before estimation or non-osmotic uptake of the sucrose may alter the concentration of solutes in the vacuole. In addition, the concentration of dissolved substances may vary considerably from day to day or during the year. For instance, many storage organs (e.g. beetroot) become considerably desiccated during storage.

Calculation of the osmotic potential as a pressure in atmospheres is possible as it is known that a molar solution of an unionised substance (e.g. sucrose) exerts a pressure of 22·4 atmospheres at 0°C. At higher temperatures the osmotic potential increases. Table 3.1 is useful for calculating water diffusion potentials.

Fig. 3.3 illustrates the relationship between cell plasmolysis and water potential of a variety of tissues. If investigations of incipient plasmolysis are carried out using ionised substances such as potassium nitrate, the solutions exert a higher osmotic potential than equivalent molarities of unionised substances. This is because the pressure exerted by the

TABLE 3.1

Osmotic potentials of different molarities of sucrose at 20°C

Molarity	*O.P.* (*atmospheres*)	*Molarity*	*O.P.* (*atmospheres*)
0·05	1·3	0·55	16·0
0·10	2·6	0·60	17·8
0·15	4·0	0·65	19·6
0·20	5·3	0·70	21·5
0·25	6·7	0·75	23·4
0·30	8·1	0·80	25·5
0·35	9·6	0·85	27·6
0·40	11·1	0·90	29·7
0·45	12·7	0·95	32·1
0·50	14·3	1·00	34·6

solutions depends on the number of particles in the solution and this is clearly higher with an ionised substance. Osmotic potential values obtained with such electrolytes will be lower than those obtained with unionised substances like sucrose.

Other methods for measuring water potential

It is possible to measure the water potential by determining weight or size changes in whole tissues. Pieces of tissue of known size or weight are placed in a range of solutions similar to those used for determining incipient plasmolysis. If there is no change in size or weight of the tissue then the water potential is equivalent to the pressure exerted by the external solution. Fig. 3.4 shows such a result with beetroot, measuring changes in length of similar pieces. There may be differences in the beetroot material used in Figs. 3.3 and 3.4, but it is clear that the water potential of the beet cells at incipient plasmolysis is much higher (i.e. more negative, about 19 atm)

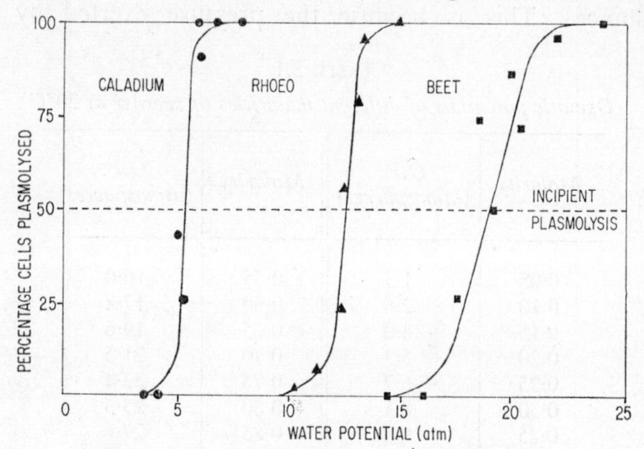

FIG. 3.3. Plasmolysis and water potential

than the water potential (about 8·0 atm) obtained by the strip method above. This would be expected from the graph, Fig. 3.2 (p. 20).

Active water uptake

The first suggestions that there might be an active or metabolically controlled means of water uptake in addition to the forms of uptakc already discussed came from an inspection of the results of different methods of measuring the osmotic potential of cells (Bennet Clark *et al.*, 1936). Values obtained by plasmolysis were higher than the readings taken by the lowering of the freezing point of expressed cell sap. Expressed sap contains some cytoplasm as well as the contents of the vacuole. The suggestion was that the cytoplasm contains a lower concentration of dissolved substances than the vacuole, ions accumulating in the vacuole by energy requiring processes. As these pass into the vacuole they increase the osmotic

PLATE 1. An alpine cushion plant (*Raoulia eximia*), sometimes referred to as the 'Vegetable Sheep', growing at 4800ft on Mount Torlesse, Canterbury, New Zealand.

Courtesy: Alpine Garden Society

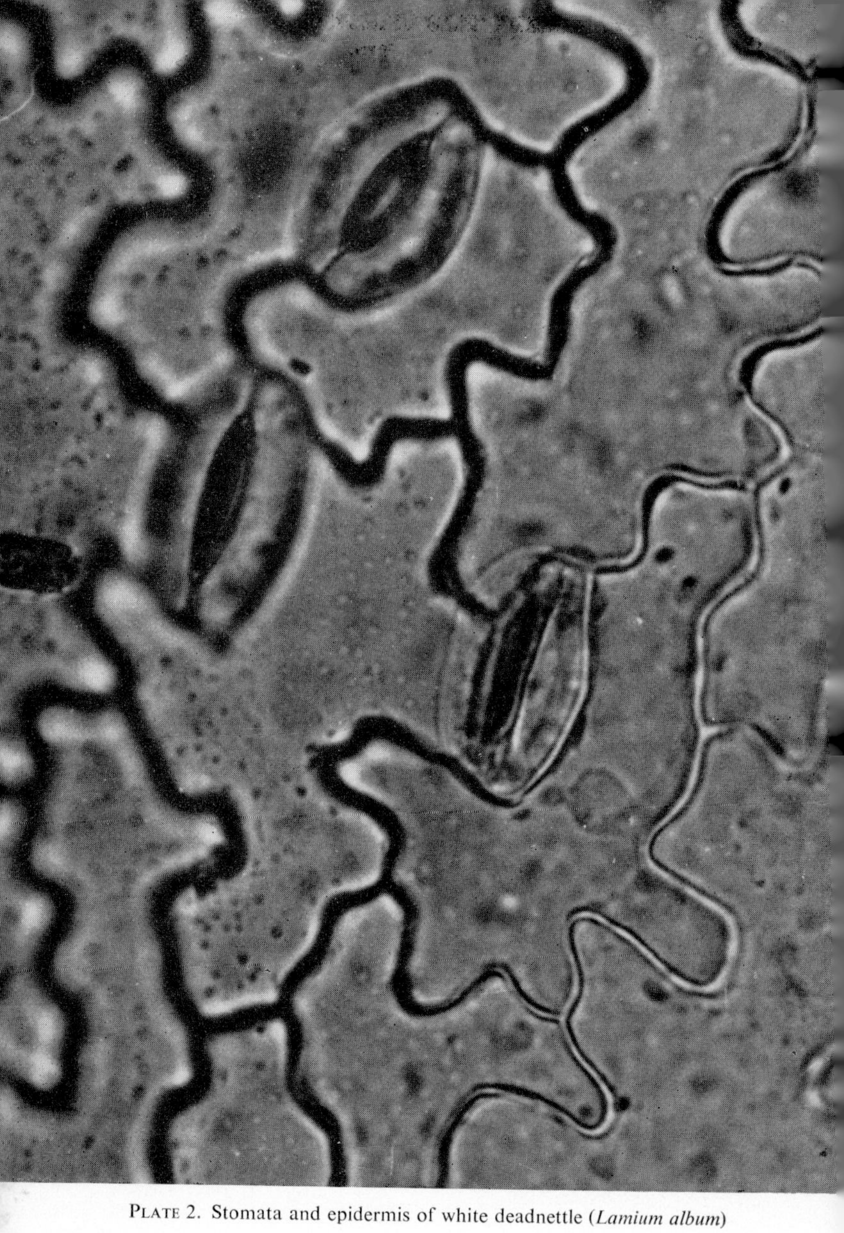

PLATE 2. Stomata and epidermis of white deadnettle (*Lamium album*)

potential of the cell sap. This will, in turn, permit a further *osmotic* uptake of water. Osmotic uptake may therefore be indirectly linked with metabolic processes.

Hackett and Thimann (1950) tried to obtain more positive information on the subject by investigating the effect of lowered oxygen concentration and respiratory inhibitors on water uptake. They placed potato discs in water and supplied them with different quantities of oxygen. At low oxygen

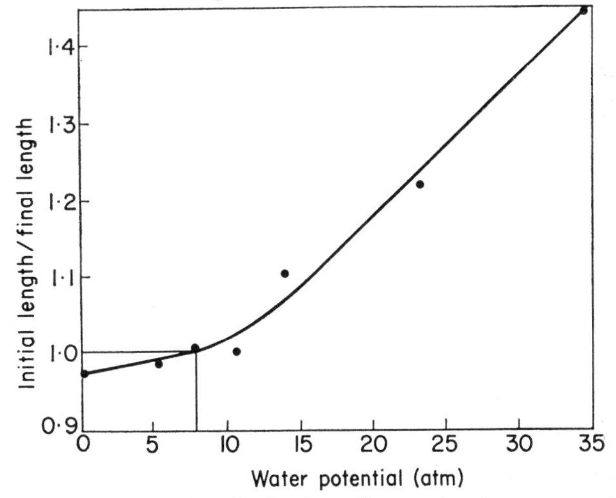

FIG. 3.4. Change in length of a tissue (beetroot) and water potential

tensions the rate of respiration and the amount of water uptake were found to be reduced.

The effect of respiratory inhibitors such as sodium azide and dinitrophenol (DNP) is more complex. Both inhibit water uptake. Sodium azide inhibits respiration at the same time, but DNP, which uncouples the oxygen used for respiration, stimulates oxygen uptake.

Other substances such as auxins (indolylacetic acid, IAA) also affect osmosis. IAA is known to influence cell elongation, probably by lowering the elasticity of the cell wall. This may reduce the wall pressure and facilitate further osmotic uptake.

Fig. 3.5 shows how the rate of water exudation (and dissolved substances) from cut stem stumps of the sunflower (*Helianthus annuus*) is affected by auxins. The cut stems were kept in continuous darkness and at a constant temperature. Two points are clear: first, that auxin addition causes a highly significant rise in the rate of exudation (which derives, presumably, from increased water uptake). Second, that there is

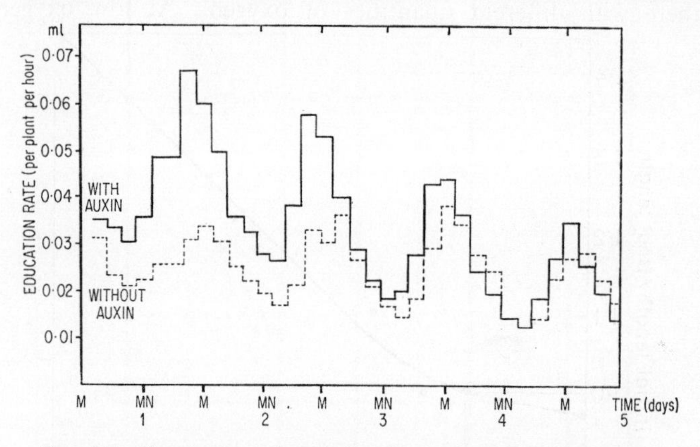

FIG. 3.5. The rhythm of exudation from cut stem stumps of the sunflower (*Helianthus annuus*) with and without the addition of auxin

a strong daily periodicity in the exudation. This is one of the fascinating cases in which the plant appears to possess a 'built-in' rhythm.

There are then a number of metabolic factors affecting water uptake into the cell. It seems doubtful, however, if their effects are particularly direct and it is probable that the bulk of water uptake into the cells at least is by the purely physical process of osmosis.

4

Water Loss

Introduction

Over 90 per cent of water uptake by the plant is lost by evaporation from the plant surface. Most of the energy needed for this uptake is provided by the heat of the sun causing evaporation rather than by metabolic processes. This loss of water by evaporation is called *transpiration.* In addition plants may in some cases actually exude water; this exudation may be partly osmotic and partly metabolic in nature and is called *guttation.* Apart from exudation caused by damage these are the only means by which plants lose water. The quantity of water lost in transpiration may be as much as a ton per hour from a large tree. Consequently, transpiration is one of the most important phenomena in plant physiology.

The significance of the role of transpiration

Gaseous diffusion from cell surfaces. If the process of respiration is to go on, most plant cells must be supplied with oxygen and able to lose carbon dioxide. Conversely, if the plant is to make its net gain in carbon through photosynthesis, it must be able to take in carbon dioxide from the atmosphere and lose oxygen. Free gaseous diffusion is therefore a necessity to all green plants.

The obvious snag is that if these gases are to be exchanged freely with the atmosphere, the cells concerned are almost bound to lose water vapour at the same time. The only exceptional case would be if the atmosphere were very nearly saturated with water vapour, but this is a rare occurrence under natural conditions. The evaporating surface is an involved one. The principal cell tissues concerned are the mesophyll cells of the leaf. These form a loosely packed, spongy tissue. Mesophyll cell walls have a fairly simple structure, with cellulose and calcium pectates embedded in a system of microfibrils, surrounding the watery cytoplasm and vacuole. The microfibrils are between 100 and 300 Å in diameter and are separated by water spaces of a similar size which form a considerable part of the cell wall volume. Water molecules are about 1·9 Å in diameter. There is plenty of room for them in between the microfibrils! Water, containing dissolved oxygen and carbon dioxide is thus able to diffuse freely from the cytoplasm and through the cell wall.

Cooling. As water evaporates it takes up its latent heat of vaporisation and the evaporating surface is cooled. This has often been quoted as being of considerable significance to plant life, as leaves can get quite hot due to absorption of the sun's radiant energy. Most of the energy absorbed will be radiant heat but some will be light energy, particularly infrared. Too high a leaf temperature could inactivate enzyme systems, but there are recorded cases of succulents being able to stand temperatures of 50–60°C without apparent damage.

In 1923 Miller and Saunders determined the leaf temperatures of various crop plants using simple thermocouples. Turgid leaves that were transpiring strongly were compared with some that had been allowed to wilt. Table 4.1 shows their results with the cow pea (*Vigna sinensis*).

TABLE 4.1

Leaf temperature and wilting

Condition	*Temperature* (°C)
Air temperature	37·0
Turgid leaf	36·5
Wilted leaf	46·0

The rate of transpiration in the cow pea plant with turgid leaves was twenty times that of the plant with wilted leaves. Two other plants, corn and sorghum, showed less marked differences, of the order of one or two degrees. It would seem unlikely that in plants such as these transpiration cooling is of much significance.

The transpiration stream. The most important result of transpiration is to cause the passage of water through the plant to replace the water loss. At the same time solutes, particularly minerals that have been absorbed by the root are transported through the plant.

Transpiration has then, some positive value to the plant. It is not just a 'necessary evil'; a process that has to go on if gaseous diffusion is to take place. The fact that plants growing in a water saturated atmosphere produce strangely differentiated cells suggests that the internal effects of a rapid transpiration stream may be more beneficial than is often realised.

The measurement of transpiration

Water vapour. One of the earliest experiments on transpiration was carried out by Guettard in 1748. He enclosed a leafy shoot inside a large glass jar (see Fig. 4.1). Water vapour condensed on the inside of the containing vessel and was collected. This set-up interferes with the normal air humidity surrounding the plant and cannot be useful except as a demonstration of transpiration.

A more efficient system was developed by Freeman in 1908. His apparatus is illustrated in Fig. 4.2. Air is drawn over a transpiring shoot and the water vapour is collected in two U-tubes containing phosphorus pentoxide. These tubes are weighed and the weight increase during the experiment gives the water content of the atmosphere surrounding the shoot. A suitable control, without the shoot, must also be set up to make allowance for the water content of the atmosphere alone. In this way the quantity of water produced by the plant on its own can be measured. This apparatus is much less subject to criticism than Guettard's but the current of air passing over the shoot is rather unnatural.

FIG. 4.1. Guettard's apparatus (1748) for demonstrating transpiration

Potometers and the measurement of water uptake. The potometer method of Vesque has already been described in connection with water uptake (Fig. 2.1). Potometer methods are some of the simplest for estimating transpiration rate since water uptake and water loss are usually very similar. The value of modern potometers is largely for determining the factors that affect the rate of transpiration and for comparing the transpiration rates of similar sized shoots. Modern potometers are simpler than Vesque's. One of these is illustrated n Fig. 4.3. A bubble of air is held in the PVC connection. When the PVC is squeezed the bubble is forced into the beginning of the capillary tube. The bubble is forced back to its original

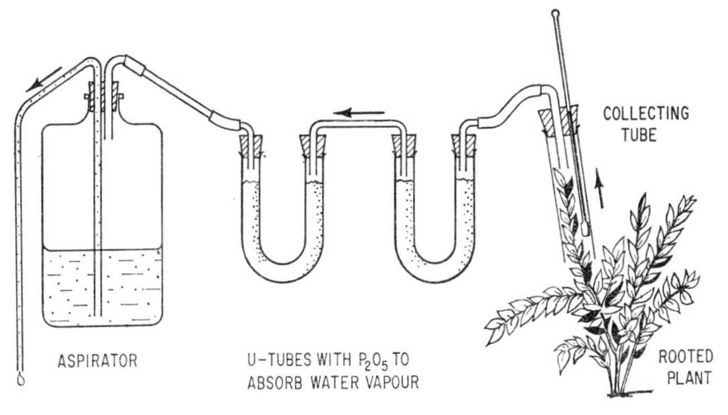

FIG. 4.2. Freeman's apparatus (1908) for estimating transpiration

position for a fresh experiment by opening the reservoir tap for a moment. The rate of water uptake is measured by noting the distance the bubble travels along the capillary in a given time. This distance is converted into a volume by measuring the bore of the capillary. The advantage of this system is that it allows the plant to continue water uptake for many hours without the danger of air-locks being created. Readings can be taken when convenient.

Indicator methods. Filter papers impregnated with cobalt chloride and cobalt thiocyanate both change colour from blue when anhydrous to red-purple when they are moist. The change in colour is gradual and a colour standard for comparison can be made up. A modern technique makes use of a system used by the building trade for estimating wall humidity. The filter papers are arranged like surgical plasters and stuck on to the upper or lower surface of the leaf. The paper is then left for at least one hour and the colour change observed.

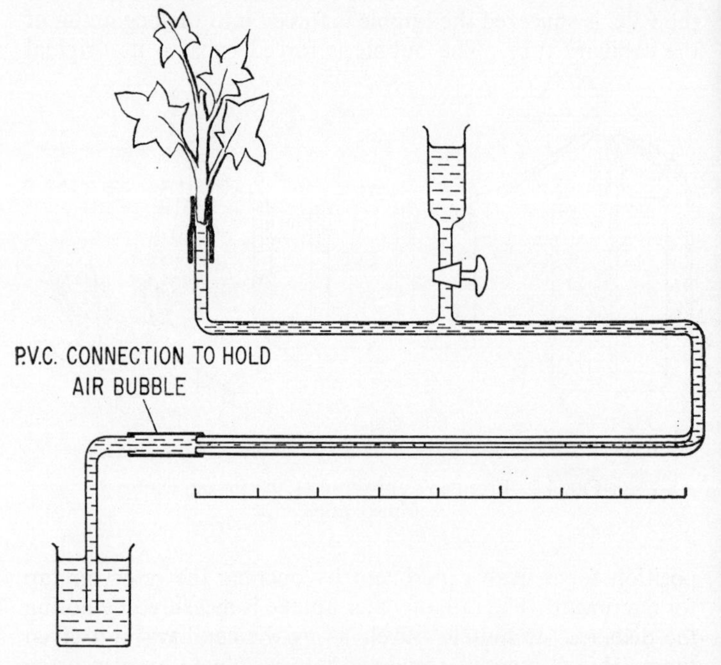

FIG. 4.3. A modern potometer

A standard chart allows for conversion to exact relative humidities.

Covering the leaf surface does of course interfere with the natural evaporation rate, but it gives a useful insight into the amount of water vapour that diffuses through the leaf surface. The results of a simple experiment of this sort are given in Fig. 4.4. The conclusion is that the lower surface of the leaf (where a high proportion of the stomata are situated) tends to be more humid than the upper surface. Second, different species give different results. This technique is useful for comparing transpiration rates of plants in the field.

Weighing methods. Loss in weight of plant material is often regarded as a gauge of water loss, since changes in water

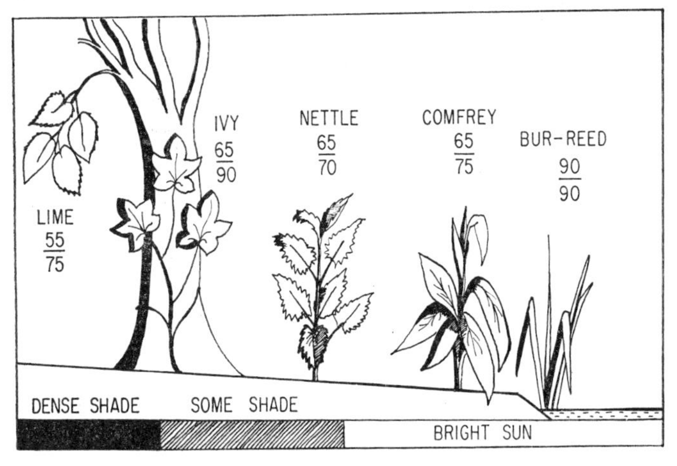

FIG. 4.4. Leaf surface humidities of a group of waterside plants

content involve greater weight changes in the plant than do changes in the content of any other substances present (e.g. carbohydrates). One weighing method has already been described with reference to water uptake. Fig. 4.5 shows how uptake and loss compare in a simple investigation of this sort. In this case cut shoots of ivy in the laboratory actually lost more water than they took up. Presumably the removal of the root system had disrupted the mechanisms for water intake and transport. Many investigators weigh the whole plant with the pot sealed to prevent evaporation. Briggs and Shantz in 1915 developed automatic recording balances capable of weighing up to 200 kg with a 5 g sensitivity. Such techniques probably give the most accurate values for transpiration loss.

Recently gravimetric methods have been simplified by Hygen (1951) and Willis and Jefferies (1963) who estimated the weight losses of detached leaves. They determined the decrease in transpiration of cut leaves by weighing them on a

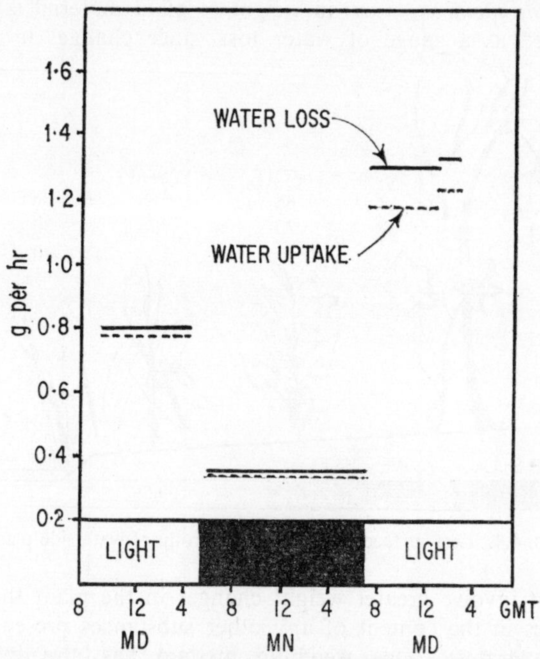

Fig. 4.5. Water uptake and water loss in cut shoots of ivy (*Hedera helix*). In this experiment water loss exceeded water uptake and shoot is therefore increasing its water deficit

torsion balance (sensitive to 0·01 mg) at frequent intervals so as to obtain a *transpiration decline curve*. Most leaves show a curve similar to that shown in Fig. 4.6. Such a curve can usually be divided into three main parts. First there is a rapid phase called the *stomatal phase*. In the second phase the transpiration decreases rapidly as the stomata close; this is the *closing phase*. Finally there is the period when water loss is slight, called the *cuticular phase*. These phases are not always particularly distinct and depend to a great extent on the time of day and the type of leaf under investigation. In spite of this, the transpiration rate at the time of cutting the leaf may be safely obtained by extrapolation from the values

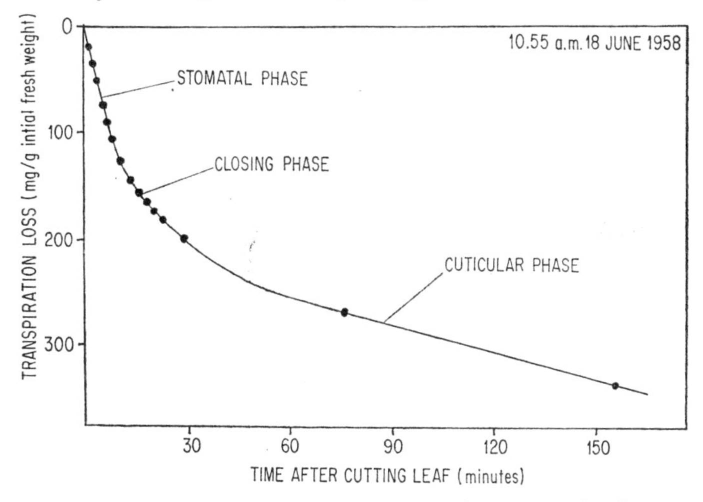

FIG. 4.6. Transpiration rate of ragwort (*Senecio jacobaea*)

obtained during the first three minutes. This method is therefore useful for accurate transpiration measurements.

Physical factors affecting the rate of transpiration

The measurement of evaporation rates. If transpiration is a purely physical evaporative process carried out without metabolic control, it would be expected to correspond closely to evaporation rates under various climatic conditions. Evaporation rates are measured by means of an atmometer. A simple version of such an instrument is identical with the potometer illustrated in Fig. 4.3 except the plant is replaced by a porous pot, or similar surface, whose area can easily be calculated.

An even simpler set-up involves the use of saturated pieces of filter paper. These can be weighed on a torsion balance at the beginning of the experiment and then at about one minute intervals for about ten minutes. This is a simple and most satisfactory method.

The physical factors. In practice it is not particularly easy to show how transpiration rates are related to any *one* physical factor. This is due to the obvious difficulty of ensuring efficient control of other environmental factors. The factors that affect evaporation rates are also those that affect transpiration; these are humidity, temperature and wind velocity. Soil conditions and the plant's internal water balance are also important in their effect on transpiration rates, as is light in influencing the opening and closing of the stomata.

Humidity. Transpiration and evaporation both show a linear relationship to the percentage saturation of the atmosphere: *the relative humidity.* The results of work by Martin in 1943 using the sunflower (*Helianthus annuus*) illustrate this relationship. The experiment was carried out with the plants at a controlled temperature in the dark, but it was arranged that their stomata were open (see Fig. 4.7).

Temperature. Water vapour pressure increases with temperature. Consequently the higher the temperature, the higher will be both evaporation and transpiration rates. The result of a simple experiment using a potometer at different temperature conditions is illustrated in Fig. 4.8.

Wind velocity. As would be expected, increased wind velocity removes water molecules from the neighbourhood of evaporating surfaces and so, at least in the normal velocity range, increases both transpiration and evaporation rates. Physical shaking of the leaf by the wind causes the compression of the intercellular spaces and will also increase the rate of water loss from the leaf. The results of an experiment by Martin and Clements in 1935 are shown in Fig. 4.9. At all velocities wind has a profound effect on the rate of transpiration of the sunflower.

The water deficit of the leaf. An internal factor that influences transpiration rate is the water deficit of the leaf. The water deficit is defined as the amount of water required by the leaf to reach full turgor, and it is expressed as a percentage of the water content of the fully turgid leaf. It depends on the external factor of the availability of water in the soil and on the internal one of the rate of absorption, transport, and utilisation of water within the plant. Changes in the water

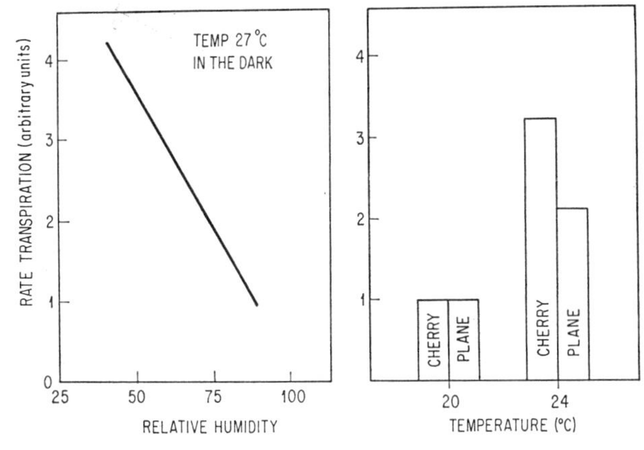

FIG. 4.7. Relative humidity and transpiration in sunflower (*Helianthus annuus*)

FIG. 4.8. Temperature and transpiration in cherry and plane tree

deficit of a leaf have a profound effect on the stomata. As the deficit increases, the stomata may start to close, thus reducing the transpiration rate. The experiment of Chung (1935) over three days show clearly that soil desiccation causes transpiration rates of the runner bean (*Phaseolus vulgaris*) to decline as the plant's water deficit increases (Fig. 4.10).

Light. Light in itself has no effect on evaporation rates (though the absorption of some wavelengths, e.g. infra-red may result in a heating of the leaf), yet its effect on transpiration is most important. Light usually causes the opening of the stomata within a few minutes and this allows for greatly increased water loss.

The diurnal fluctuations in transpiration. Transpiration almost always shows a clear diurnal rhythm, rising to a maximum during the day and falling off at night. Simultaneous, accurate determinations of the various factors described above show

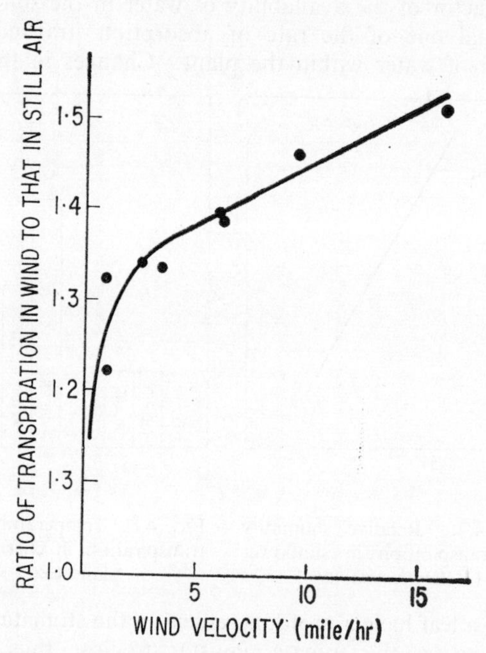

FIG. 4.9. Wind velocity and transpiration in sunflower (*Helianthus annuus*)

that in some cases the correlation with evaporation rates is not as close as might be expected. Willis and Jefferies (1963) took transpiration measurements by the torsion balance weighing method described above, on various plants including the hound's-tongue (*Cynoglossum officinale*) growing on the dry slopes at Braunton Burrows in North Devon. In addition, they measured leaf water deficit, air temperature, relative humidity, light intensity and stomatal aperture. The hound's-tongue showed a typical diurnal transpiration curve with the peak at about 10 a.m. The interesting point was that transpiration began to decline *before* maximum evaporation rate was

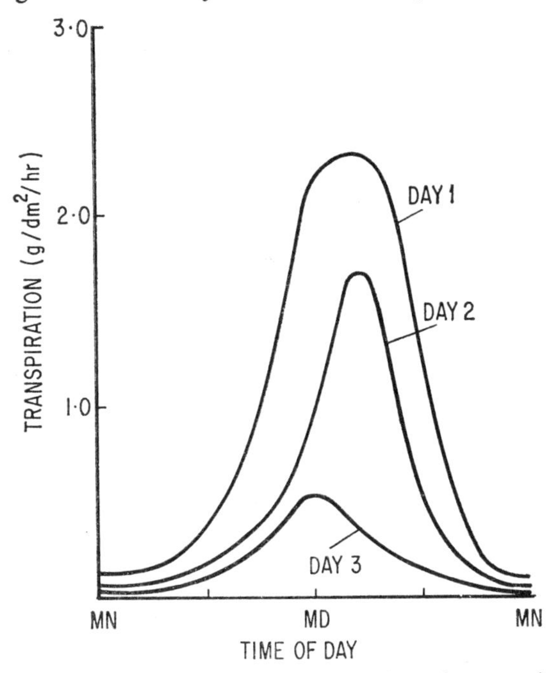

FIG. 4.10. Diurnal rhythm of transpiration of runner bean (*Phaseolus vulgaris*) as the soil becomes progressively drier over three days

reached (see Fig. 4.11). The plant's water deficit also continued to rise after the time of maximum transpiration had passed. This is to be expected in view of the dry habitat of the hound's-tongue from the discussion of Livingston's findings at the end of Chapter 2.

The factor that showed the most direct correlation with transpiration was, perhaps hardly surprisingly, the stomatal

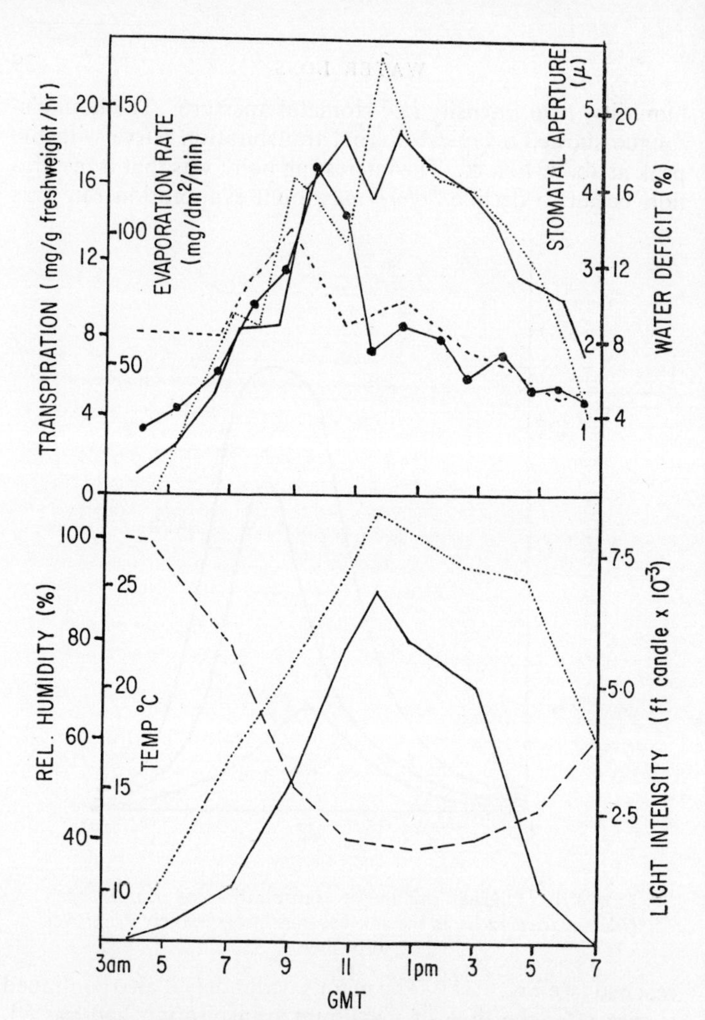

FIG. 4.11. Diurnal changes in transpiration of Hound's tongue (*Cynoglossum officinale*). Upper part of the figure shows the transpiration rate —·—·—·—, evaporation rate ———, stomatal aperture – – –, and water deficit ·········. The lower part shows light intensity ———, humidity ------ and air temperature ·········.

aperture. These structures must therefore rate as one of the most important regulatory mechanisms in plants.

Morphological and anatomical features affecting transpiration

Growth form. If a plant is to survive and grow it must have the right structural and physiological balance with respect to the environment in which it is found. It must be able to obtain sufficient carbon dioxide to make a net daily gain in organic material and yet must not lose so much water that its cell processes are jeopardised.

The existence of global vegetation zones, each characterised by a range of species, many of which have a similar growth form, suggests that certain clearly defined structures are best suited to survival in a given environment. Of all the factors of the environment, probably water content of the soil and atmosphere are the most important in determining the selection and survival of different growth forms. The cushion-shaped mountain plants (see Plate 1), deciduous forests and the sclerophyllous vegetation of the Mediterranean are all forms which are, at least to some degree, reflections of water availability. Even the bulbous plants that are common in Mediterranean countries are adapted for withstanding *summer* drought after completing most of their growth during the wetter winter months.

Surface:volume problems. Like so many other problems in biology, growth form is subject to surface:volume ratio considerations. The higher the ratio the greater will be the transpiring surface and *vice versa.* Photosynthetic efficiency will be greater if the surface area through which light may be absorbed is increased. The result is that selection in favourable environments has favoured the survival of plants with a rather diffuse structure and a high surface to volume ratio. It is generally in the succulents and xerophytes that low ratios are to be found; these plants may have specialised photosynthetic zones to make up for the lack of the more usual leafy tissue. Most plants, therefore, depend for water conservation on specialised features, both at a microscopic and even sub-microscopic level. Such adaptations will be found mostly in the outside layers of the stem and leaf; in the epidermis.

The leaf epidermis and cuticular water loss. In most plants there is an upper and lower epidermis, each consisting of one layer of cells. The epidermal cells are usually more or less transparent, containing few chloroplasts and their main function is for protection, particularly against water loss. The experiment described on p. 32 shows that much more water vapour diffuses through the lower surface than the upper surface of a leaf. This is usually the case in dicotyledons, but in many monocotyledons (such as the bur-reed, *Sparganium erectum*, see Fig. 4.4) similar quantities may diffuse through both surfaces. This diffusion must take place through the stomata and the outer lining of the epidermis, *the cuticle* (see Plate 2). Even though these leaf surface humidities seem related to stomatal distribution and density, Fig. 4.6 shows that the loss of water through the cuticle may be quite considerable, even as much as 50 mg/g fresh weight per hour.

The cuticle is a thin deposit of waxes and fatty derivatives such as cutin and suberin on the outside of the epidermis and it is usually thicker on the upper surface of the leaf. Inspection of the leaves and stems of plants with a blue, waxy bloom like the candle plant (*Kleinia articulata*) and some spruce trees, may suggest that they have a highly impervious covering. Electron microscope photographs show however that the waxy surface is anything but uniform. Plates 3*a* and 3*b* show the surfaces of these plants; the spaces between the waxy deposits must allow for considerable gaseous diffusion. Be this as it may, the cuticle plays a role secondary only to the stomata in the successful water economy of a plant.

Guttation

Under conditions of high air and soil humidity, when rates of transpiration are low, plants often exude water from the edges of their leaves. This process is known as *guttation* and often occurs in greenhouses where drops of water may be found surrounding the leaves of tomatoes, begonias, campanulas and a wide range of plants (see Plate 4*a*).

Such plants possess water-secreting glands known as *hydathodes*. Fig. 4.12 shows a leaf section of *Saxifraga*. It can be seen how the xylem elements connect up with a group of rather loosely packed cells called the epithem, below the

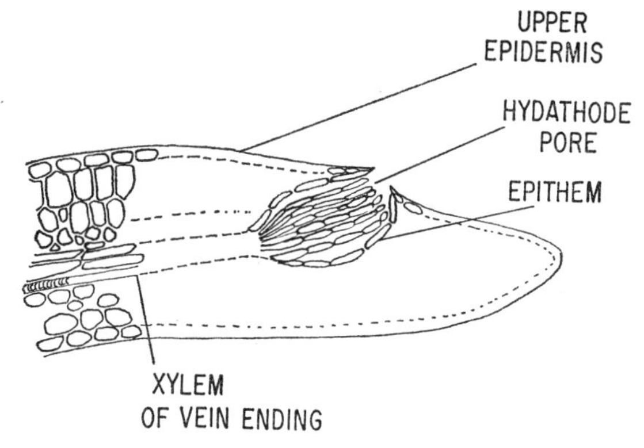

FIG. 4.12. Hydathode of *Saxifraga* leaf (highly magnified)

hydathode itself. In other cases the stomata themselves probably act as hydathodes.

Guttation probably permits the movement of water and dissolved substances through the plant when movement due to transpiration is negligible. It appears to be similar to root pressure in that its rate is affected by metabolic processes linked with respiration. Some species of saxifrage in the so-called encrusted group (*Euaizoonia*) have white deposits of salts around their hydathodes. This is due to evaporation of secreted water leaving a deposit of salts originally contained in solution. Sometimes, re-absorption of highly concentrated solutions may take place and this results in osmotic burning of the leaf margin.

The quantity of water exuded by a guttating plant may be prodigious. One of the highest recorded cases is the taro or coco (*Colocasia antiquorum*). This is an Indian plant of moist and shady habitats, with leaves about a metre and a half long and over half a metre wide. It is recorded as producing as much as 200 ml. from a single leaf in one day!

The wide range of occurrence of guttation suggests that it may be a process of much greater significance in the plant's water relations than is often realised.

5

The Stomata

Distribution of the stomata

Distribution in mosses and liverworts. The development of the stomata as a means for controlling gaseous diffusion must rate as one of the important steps in the evolution of land plants. Amongst the simpler land plants, the liverworts seldom possess

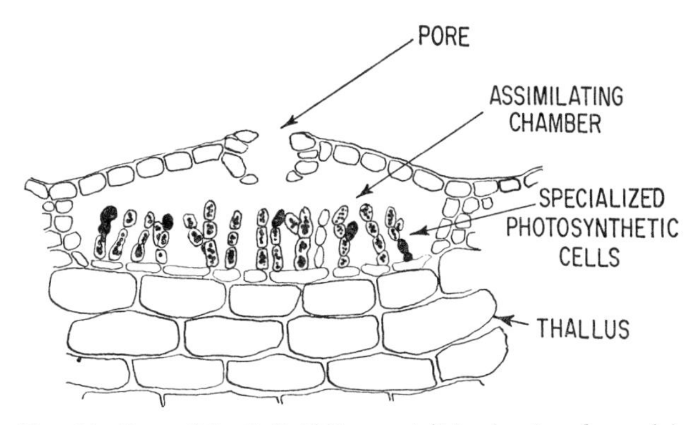

FIG. 5.1. Pore of the thalloid liverwort (*Marchantia polymorpha*) (highly magnified)

stomata at all. The more highly adapted thalloid species of liverwort, e.g. *Marchantia polymorpha* possess pores, the function of which seems to be to aid gaseous exchange. No regulatory mechanism is involved in this simple structure (see Fig. 5.1). In the mosses, and the liverwort *Anthoceros*, the capsule wall has quite well-developed stomata. These have been studied in detail by Paton and Pearce (1957) who found that bryophyte stomata showed no diurnal rhythm. They were opened by higher atmospheric humidities, but were light insensitive, and also insensitive to carbon dioxide concentration, in marked contrast to stomata of the higher plants.

The mechanism here seems to depend solely on turgor changes in the guard cells. The active life of moss capsule stomata is usually fairly short as the older stomata often become plugged with wax.

Distribution of stomata in vascular plants. Two simple methods have been used for studying the distribution of stomata in vascular plants. In the simpler technique, an epidermal strip is made and examined under the microscope. Another method, suitable where it is difficult or uneconomical to make an epidermal strip is to make a replica of the surface in cellulose. This method is particularly useful on cacti and succulents. Ordinary transparent nail varnish is painted on to the surface of the plant. After a few minutes the film is removed and examined under the microscope. The results are often quite as satisfactory as an ordinary epidermal strip.

Work of this sort reveals that the distribution, number, and size of the stomata is extremely variable; Table 5.1 gives information from several different species.

There may also be considerable variation within the same plant. These differences may be related to the position of the leaf on the plant or to the conditions (e.g. light or humidity) under which the plant is growing. Table 5.2 shows how the leaf position affects stomatal counts in the cocksfoot grass (*Dactylis glomerata*).

Plants growing in the sun and under drier conditions tend to have fewer stomata than specimens of the same species growing in moister, shady conditions. The size, shape, and density of ordinary epidermal cells also vary with the environmental conditions though it is worth noting that the ratio of the number of stomata to the number of epidermal cells remains fairly constant for a particular species. Apart from the actual numbers of stomata, the size of the stomatal pore must be an important factor determining the rate of gaseous

TABLE 5.1

The distribution and size of stomata on the leaves of different species

Species	*Average number of stomata per sq. mm*		*Size of fully open pore* (μ)	
	Upper epidermis	*Lower epidermis*	*Length*	*Breadth*
Runner bean (*Phaseolus vulgaris*)	40	281	7	3
Castor oil plant (*Ricinus communis*)	64	176	10	4
English ivy (*Hedera helix*)	0	158	11	4
Geranium (*Pelargonium* sp.)	19	59	19	12
Maize (*Zea mais*)	52	68	19	5
Nasturtium (*Tropaeolum majus*)	0	130	12	6
Oat (*Avena sativa*)	25	23	38	8
Sunflower (*Helianthus annuus*)	85	156	22	8

TABLE 5.2

The relationship between stomatal density and the position of the leaf on the plant (cocksfoot, Dactylis glomerata)

Tier of leaf	*Height of its insertion above ground* (cm)	*Number of stomata per field of view* (*Upper epidermis*)
1	0	34
3	10·2	42
5	25·2	61
6	37·0	80
7	51·0	64

diffusion through the stomata. The size of the pore may vary considerably, but is seldom more than a few microns across (see Table 5.1). To understand how stomatal density and pore size influence gaseous diffusion it is necessary to examine the process of diffusion.

The diffusion of gases through the stomata

The large quantities of water lost from leaves and the minuteness of the stomata led Brown and Escombe, working at Kew Gardens in 1900, to examine the problem in detail. Their findings agreed with previous work that the rates of gaseous diffusion through a single small hole was proportional to the diameter, not the area, of the opening. They investigated the diffusion of both carbon dioxide and water vapour. In one simple experiment a known quantity of sodium hydroxide was placed in a series of small tubes each covered by a septum pierced with a hole of given size. The carbon dioxide diffusing in from the atmosphere in a given time was then estimated. The graph Fig. 5.2 shows that there is a direct proportionality between the rate of gaseous diffusion and the diameter of the pore.

Brown and Escombe and later workers (Sayre, 1926) tried constructing septa with numerous holes. In one simple experiment, three tubes, each containing a known quantity of water were covered by a septum with one, two, or four holes respectively. The size of each pore was such that their total area per septum was similar in each case. The water lost in two days was then determined by weighing. Table 5.3 summarises the results of one such experiment.

These results confirm that water loss through the pores is proportional to the diameter of the pores rather than their area. Various workers have tried to account for this and it has been suggested that the more general distribution of large numbers of small pores allows for a freer, and therefore more rapid diffusion than could take place through a smaller number of larger holes.

All these artificial pore systems have of course very much larger and further separated pores than occur in the living leaf. The next step then was to find what happened when the holes were placed closer together. Weishaupt in 1935 determined

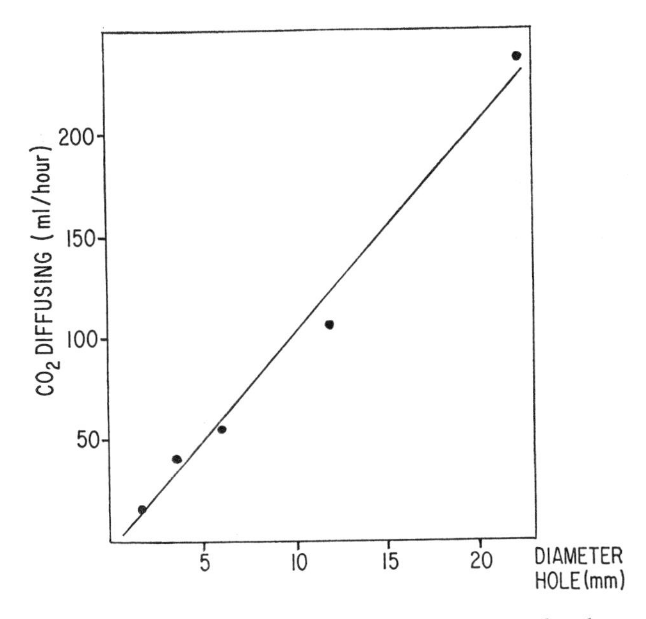

FIG. 5.2. Relation between pore size and the rate of carbon dioxide diffusion

the effect of the distance between the holes (each 0·3 mm diameter) on the diffusion capacity of the septum. Fig. 5.3 shows how the quantity of water vapour diffusing through each pore is reduced with closer packing of the pores.

The maximum diffusion capacity per pore is reached when the pores are about twenty diameters apart.

Clearly some sort of interference must take place when the pores are particularly close together. Molecules diffusing through the pores may collide with those diffusing through adjacent pores and also competition may take place between internal currents or diffusion gradients to lower the net rate of diffusion. This is shown diagrammatically in Fig. 5.4.

Brown and Escombe realised that it should be possible to devise a septum which would allow water vapour to diffuse through nearly as fast as if there was no septum present. They constructed one containing 100 holes per sq. cm, each hole

TABLE 5.3

The relationship between total diameter of pores and amount of water loss

Number of holes per septum	*Diameter of each hole* (mm)	*Total diameter* (mm)	*Total area* (sq. mm)	*Water loss* (g per 48 hours)
1	0·55	0·55	0·23	0·11
2	0·36	0·72	0·20	0·17
4	0·29	1·16	0·25	0·23

being 0·38 mm diameter. Although 88·66 per cent of the surface was still covered by the septum, water loss was very nearly the same as when the septum was removed.

They went on to compare these artificial systems with those present in living plants. Working with plants of sunflower (*Helianthus annuus*), they found that water loss was much less than the theoretically expected value obtained from work with perforated septa described above. The anatomy of the leaf in the neighbourhood of the stomata plainly offers considerable resistance to the free diffusion of water vapour.

The physical factors that influence the diffusion of water vapour through and away from the actual leaf surface were studied by Stalfelt (1932) and Syrett (1958). They summarised the factors producing resistance to diffusion of water vapour from the leaf as follows: One (R_1) type of resistance is that due to diffusion through the stomatal pores and is dependent on the number, size, and shape of the stomata. The other (R_2) is the resistance to the movement of water molecules away from the actual surface of the leaf. The component (R_2) is,

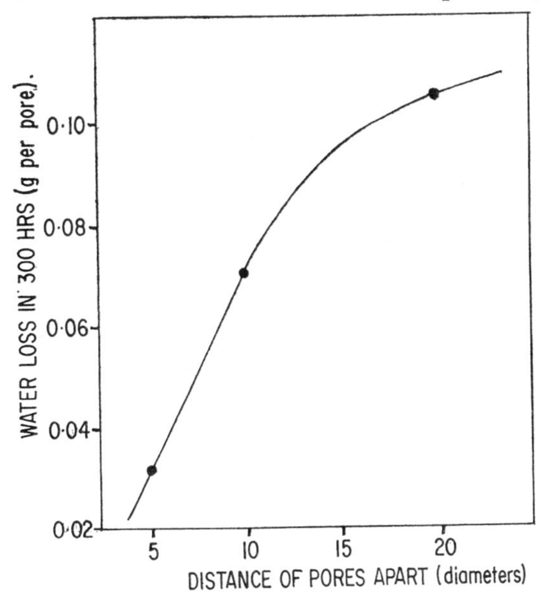

FIG. 5.3. Relation between water loss and the distance apart of the pores

in most plants, much more susceptible to wind and other external factors. In strong wind the water molecules will be moved away and the resistance (R_1) of the stomata will be relatively important though R_2 may still be greater than R_1. Plants with hairy leaves and sunken stomata have higher values for R_2 and lose less water under windy conditions. Under still air conditions the value of R_2 increases and the significance of stomatal resistance to water loss becomes relatively slight.

The structure of the stomata

Each stoma is composed of two guard cells (see Plate 4*b*) that are derived from the epidermal layer; these enclose the stomatal pore. In addition, there may be a number of epidermal cells that have become modified in association with the guard cells; these are called *accessory cells*. They are common in grasses and some other monocotyledons. The guard cells, unlike most other epidermal cells, contain chloroplasts and

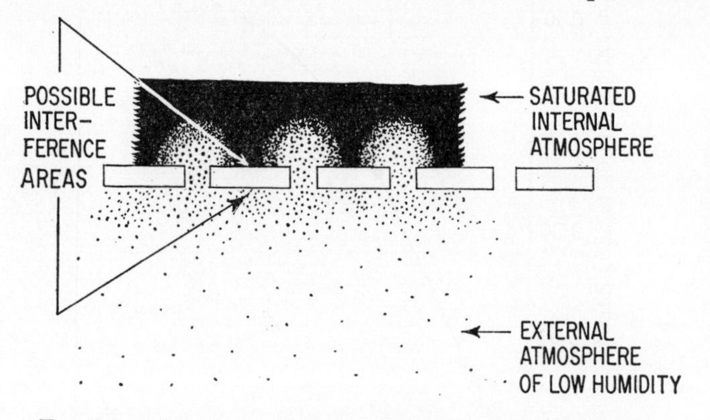

FIG. 5.4. Diagram to show the diffusion of water molecules through a perforated septum

other components like typical vacuolated plant cells. They differ from these principally in their cell walls which have characteristic thickenings, which varies in distribution with the species. In most flowering plants it is the inner wall, surrounding the pore, which is thickened and relatively inelastic. The opposite wall, on the outer edge of the guard cell, is thin and correspondingly elastic. The type of stomata found in the grasses is very different. Each guard cell is dumb-bell shaped. It is the central walls here that are extremely thick while the bulbous parts are thin-walled (see Fig. 5.5).

The arrangement of the guard cells relative to the epidermis, hypodermis and mesophyll is variable. In some cases the

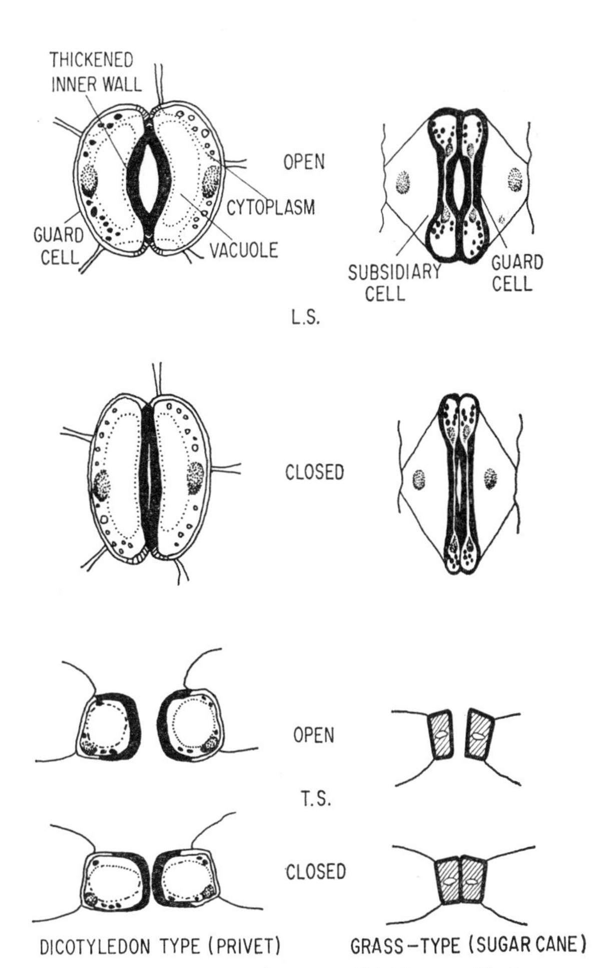

FIG. 5.5. Views of stomata (highly magnified)

guard cells lie flush with the epidermal cells, in other cases the epidermal cells overlap the guard cells on their inner edges (see Plate 2). In extreme cases the stomata may be sunk below the epidermis. This is common in conifers and plants adapted to conditions of drought.

The stomatal mechanism

The opening and closing of the stomata is one of the most obvious and important features of plant physiology yet the mechanism by which it occurs is still not properly understood. It is clear that turgor changes are responsible for the actual movement; all known stomata are open under conditions of high guard cell turgor and close when this diminishes. In the dicotyledons, where the thick inner wall is relatively inelastic, the bow-shape of the guard cell bulges out further with increasing turgidity, thus widening the pore. This phenomenon is easily demonstrated by placing an epidermal strip in a strong solution (e.g. molar potassium nitrate). This causes plasmolysis, reduction in the turgor pressure of the guard cell and consequent closing of stomatal pores. Placing the tissue in water increases the turgidity of the guard cells and the pore opens.

The stomata of the grasses open when the turgor pressure of the guard cells increase, causing a rounding-off of their bulbous ends. These ends push against one another as their size increases and so their central parts come apart (see Fig. 5.5). Some of the methods that are available for studying stomatal aperture are enumerated below.

Methods for investigating the movement of the stomata

The porometer. One of the simplest devices for investigating the mass flow of gas through the leaf is the *porometer*. This was first devised by Francis Darwin and Pertz in 1911. A version of the porometer is shown in Fig. 5.6. The cup may be fitted to the leaf by means of Vaseline or a gelatine washer.

The rate of flow of gas through the leaf is determined by taking the time for a water column in the centre tube to fall a given distance. The results of an experiment using a simple porometer are illustrated in Fig. 5.7. It has been suggested that the fitting of a cup over the leaf affects the carbon dioxide concentration around the leaf which may affect the opening

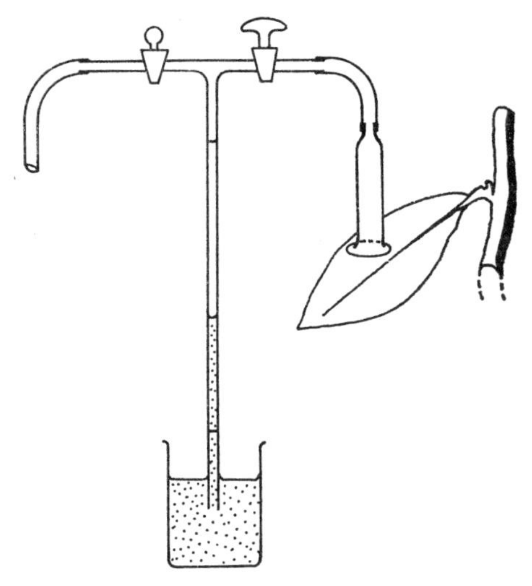

FIG. 5.6. A simple porometer

of the pore. In the light, the carbon dioxide concentration inside the cup may fall well below the atmospheric value of 0·03 per cent to as little as 0·01 per cent. Mansfield, Heath and Meidner at Reading have therefore developed porometers where the cup is in contact with the leaf for the minimal period of time. A simple porometer of this sort is illustrated in Fig. 5.8. This is basically a transparent plastic clip that is easily attached to the leaf. A reading is obtained by first removing the pipette, squeezing the rubber bulb and then inserting it into the end of the porometer clip. The bulb is then released and the time taken for it to inflate recorded. Although this sounds a little crude, it is in fact very satisfactory as results are easily obtained. Indeed it is possible to walk about in the field and

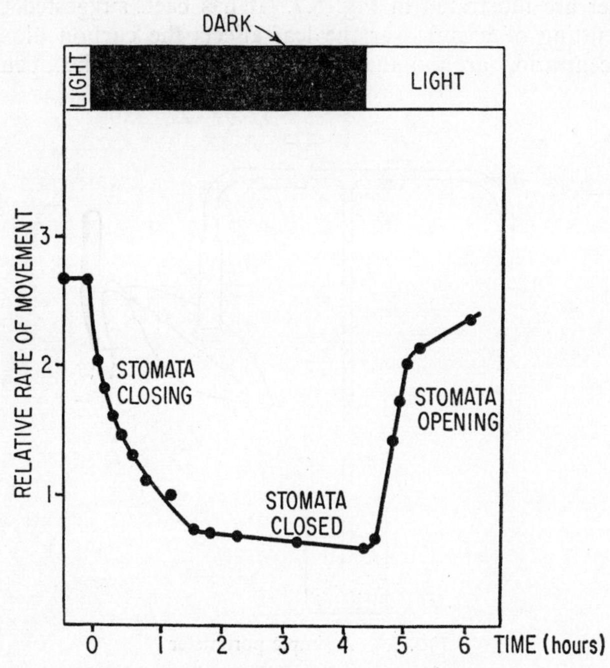

FIG. 5.7. Effect of illumination on the movement of gas through the leaf surface of *Pelargonium*

in a few seconds determine whether a plant's stomata are open or closed.

Nevertheless, the results obtained with porometers must always be treated with caution. The mass flow of gas through the leaf in this way is anything but natural. It may well affect internal factors, such as the state of the air spaces in the mesophyll. In oak, maple, and beech the veins separate the mesophyll into small discrete blocks, with relatively few air

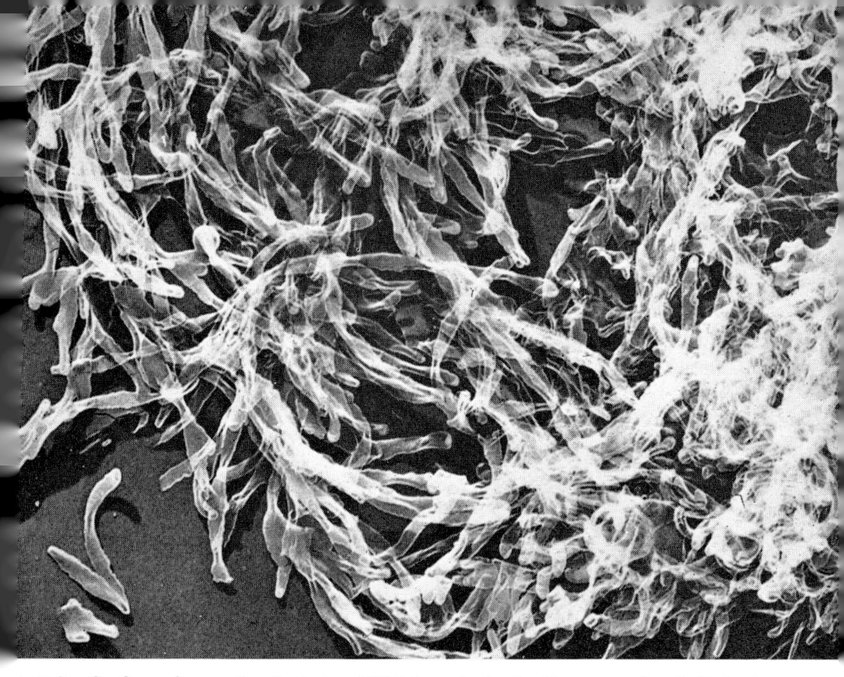

LATE 3a. Surface of a candle plant stem (*Kleinia articulata*). *Courtesy: Dr. B. E. Juniper*
Electron microscope photograph ×15,000

PLATE 3b. Surface of a spruce needle. *Courtesy: Dr. B. E. Juniper*
Electron microscope photograph ×11,000

PLATE 4a. Guttation in lady's mantle (*Alchemilla*)

PLATE 4b. Stoma of spiderwort (*Tradescantia virginiana*)

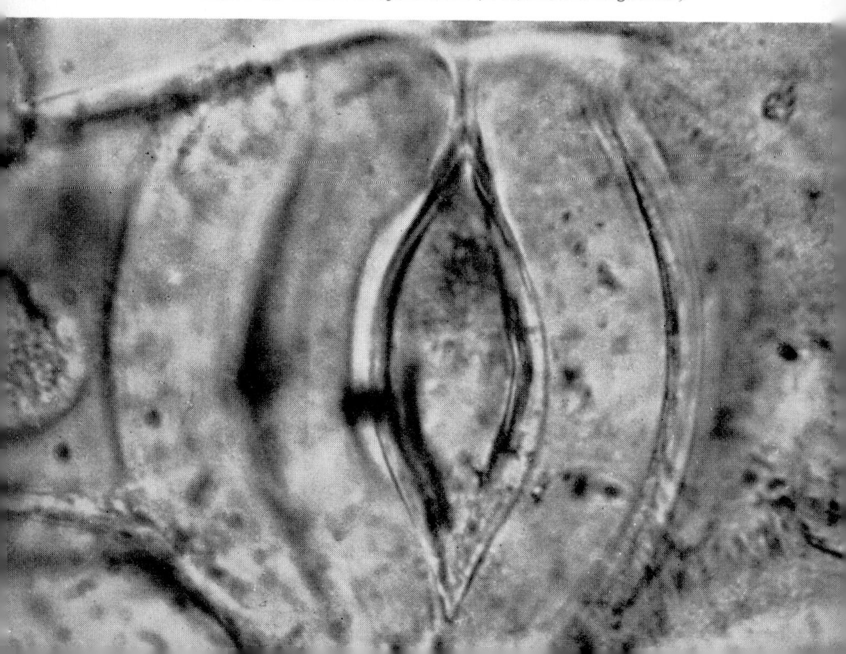

passages linking them. Such species will obviously be unsuitable for work with porometers.

The infiltration method. This method is again quick and convenient for use in the field. The method was originally

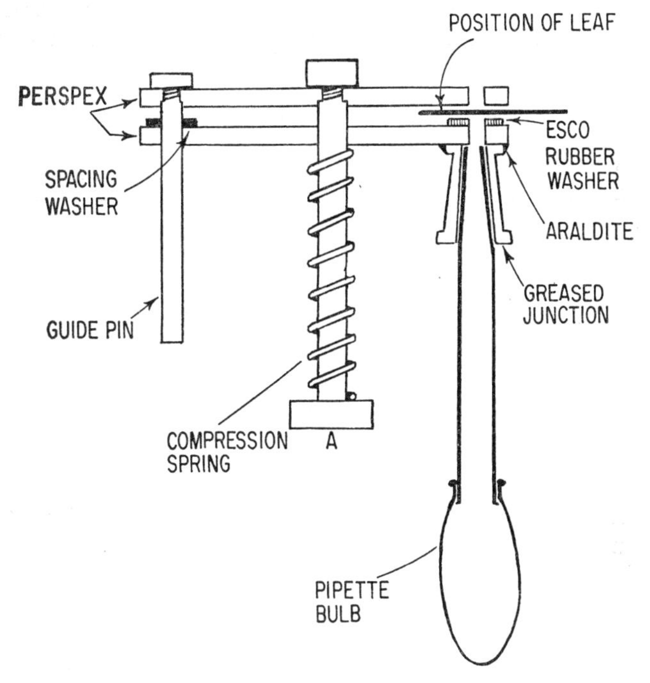

FIG. 5.8. Meidener's clip porometer. Pressure with the thumb at A and fingers on the lower piece of Perspex opens the clip so that it may be attached to the leaf.

devised by Molisch in 1912, but the methods described below have been used with great success in recent work by Geyer (1963) and Willis and Jefferies (1963).

When many organic liquids are placed on the underside of leaves they penetrate the stomata and a bright green colour results. By using substances of different viscosities it is possible to estimate the degree of opening of the stomata. Geyer used a mixture of *iso*-butanol and glycerin; ethylene glycol and *iso*-butanol could also be used and is less sticky to handle. Willis and Jefferies used mixtures of petroleum ether and ethanol, but these have the disadvantage that they evaporate quickly. Table 5.4 shows how their mixtures were made up.

TABLE 5.4

Solutions used to determine the infiltration index

Infiltration Index	1	2	3	4	5	6	7	8	9
A. Parts *iso*-butanol	9	8	7	6	5	4	3	2	1
Parts glycerin	1	2	3	4	5	6	7	8	9
B. Parts petrol. ether	10	8	6	5	2	0	0		
Parts Abs. ethanol	0	2	3	5	8	10	10*		

Solutions 1 and 2 have waxoline blue and 3 to 7 have crystal violet 100 mg/150 ml added to make assessment of penetration easier.

* 95 per cent ethanol.

Glycerin (or ethanol) penetrate less readily and so a high infiltration index indicates that the stomata are well open.

The method is quick and efficient and is suitable for most plants except those that are excessively hairy or very young. It is important to ensure that only small drops of the solutions are used otherwise air may have difficulty in escaping from the surrounding stomata. Also the distribution of the stomata may not be even, so several separate drops should be applied. Willis and Jefferies have compared their infiltration indices with measurements of stomatal apertures obtained from alcohol-fixed epidermal strips. They found that with the ragwort (*Senecio jacobaea*) there was nearly a direct relationship between the two (see Fig. 5.9).

Diurnal rhythms and stomatal movement

Geyer's work on the oleander (*Nerium oleander*) and maize (*Zea mais*) showed how useful the infiltration technique could be. He showed that while both these plants opened their stomata more or less simultaneously early in the day, maize normally closed them by about mid-day while those of oleander remained open until the evening (see Fig. 5.10). These diurnal

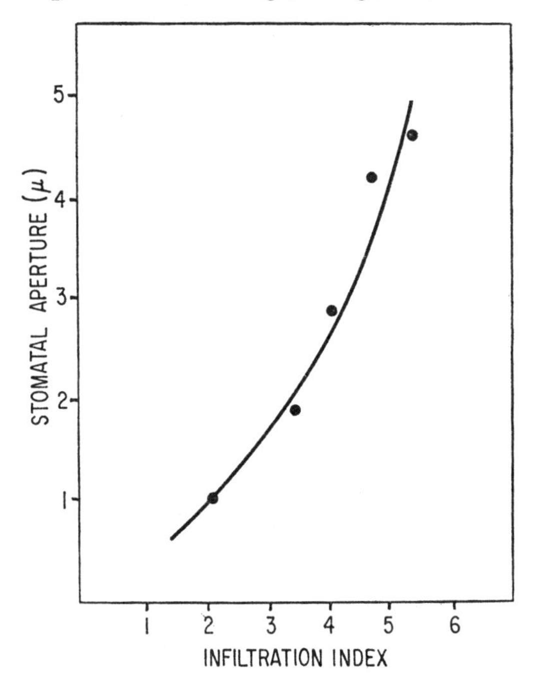

FIG. 5.9. Relation between infiltration index and stomatal aperture in ragwort (*Senecio jacobaea*)

changes in stomatal aperture are principally affected by conditions favouring high evaporation rates and by light intensities working on the stomata, but there are still a number of important and unsolved problems relating to the control of stomatal rhythms. For instance an inverted rhythm is reported to occur in many succulent plants like *Mesembryanthemum*. There are also several recorded cases of the diurnal rhythm persisting even when the environmental conditions are entirely different from normal. For instance, the plant could be in continuous light.

In other cases night opening of the stomata may take place for a short time. Recently Schwaber has found that night opening in the succulent *Kalanchoe blossfeldiana* is controlled

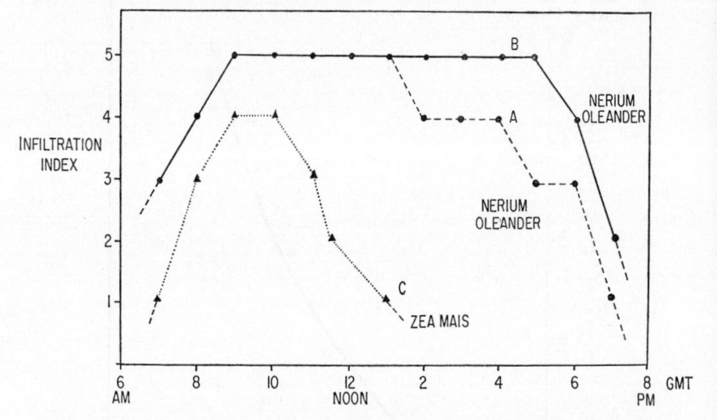

FIG. 5.10. Stomatal movement in maize and oleander. A was illuminated for the first half of the day, B and C were illuminated throughout the day

by the length of the dark period. Under short-day conditions with eight hours daylight, the stomata began to open two to three hours *before dawn*. When the dark period was shortened so as to simulate long-day conditions, no night opening took place. In addition the length of the dark period greatly influenced the rate at which the stomata opened; the longer the night the quicker they opened.

Stomata and wilting

The above discussion is based on the assumption that the stomata will open and close provided the plant has an adequate water supply. When a water deficit is beginning to build up there are numerous indications that plants close their stomata. This closure may take place either before or after the onset of wilting. The closing of the stomata of maize, illustrated in Fig. 5.10, could well indicate such a situation though this plant normally closes its stomata at mid-day. Further mention may be made of the careful, quantitative work of Willis and Jefferies on hound's-tongue, which began to close its stomata at about 9 a.m. (see Fig. 4.11). It seems possible that the rising water deficit of the plant had become critical. Closure of the stomata before maximum light intensity and evaporation rates had been reached thus prevented the danger of serious wilting in a relatively dry habitat.

Dune-slack plants, growing in an area where water is much more readily available, show very different results. Many, though not all plants in this locality keep their stomata open longer and build up less of a water deficit.

Wilting and stomatal movement have been studied further by Geyer, who carried out a simple experiment on freshly cut shoots. These were placed in the sun and stomatal movement and state of turgor were observed. The results of his experiment are shown in Fig. 5.11. Not all plants close their stomata either before or in the early stages of wilting; there are a number of recorded cases of plants keeping their stomata open during wilting and even a few instances of plants closing their stomata as wilting begins, only to open them again during the later stages.

Biochemical changes in the guard cell

If the stomata are stained with iodine, starch is only found in the guard cells that have been in the dark, usually for some time. Unlike most mesophyll and palisade cells, guard cells do not accumulate starch under light conditions. That two such diametrically opposite conditions can exist in cells that are only a few microns apart clearly requires some explanation. One clue as to the process involved comes from the observation, mentioned in the discussion of porometers, that if the carbon dioxide concentration around the guard cell is changed, it may affect the stomatal aperture. Lowering of the concentration of carbon dioxide causes the stomata to open particularly wide. Recently Heath (1950) has shown that, even in the light the stomata will close if the carbon dioxide concentration is sufficiently high (above 0·84 per cent). Clearly, a light-treated leaf is using carbon dioxide for photosynthesis and so

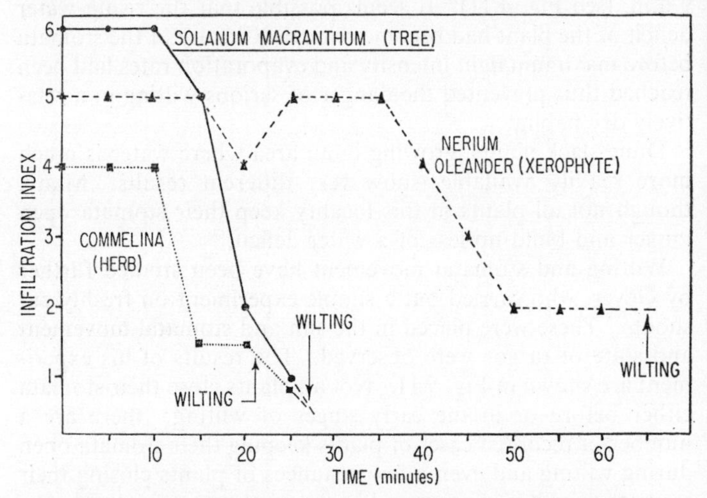

FIG. 5.11. Wilting and stomatal movement in various plants

the carbon dioxide concentration around the guard cells must be much higher during the night than in the day.

If carbon dioxide is present it will dissolve to form carbonic acid and this may alter the pH of the cell contents. Some starch forming enzymes are known to be pH sensitive, the direction of the reaction being determined by the acidity of the cell sap. Microchemical tests have indicated that sucrose is probably the main product of guard cell photosynthesis. The guard cells must lack the usual starch synthesis enzyme system but possess a special pathway of their own which can result in the conversion of sucrose to and from starch. One or more steps in this pathway must be pH sensitive. In the light, the pH of the guard cell sap rises, the starch is converted into sucrose, while the opposite occurs in the dark. A similar enzyme system has been identified in starch phosphorylase. It converts glucose-1-phosphate into a simple form of starch under acid conditions and the reaction is reversed at high pH.

Sucrose is of course osmotically active and as the osmotic potential of the cell sap rises more water will be taken up from neighbouring cells. This will increase the size of the guard cells and the stoma will open. These inter-relationships are summarised in Table 5.5.

TABLE 5.5

The conditions in the guard cell influencing the opening and closing of the stomata, in a plant which is not suffering from water shortage

Light	*Dark*
Photosynthesis and Respiration	Respiration only
CO_2 concentration low	CO_2 concentration high
pH high	pH low
Sucrose and simple sugars formed	Starch formed
High O.P.	Low O.P.
Full turgor	Less turgor
STOMA OPEN	STOMA CLOSED

Unfortunately there are many further complications. For instance, how do the stomata in non-green parts of variegated leaves open and close? A number of plants, principally monocotyledons, do not form starch at all in their leaves: there is no indication how these operate.

6

The Movement of Water and Solutes

Introduction

The movement of water through the plant. The immense quantities of water lost by transpiration are replaced by an ascending flow of water known as sap, which moves through the xylem elements of leaf, stem, and root. The importance of the non-living parts of the stem (e.g. the xylem) in the upward transport of sap was first demonstrated in 1893 by Strasburger. He cut down a 75 year-old oak tree at ground level and used the trunk, which was twenty-two metres long, as his experimental material. He immersed it in picric acid which is lethal to living plant cells, and found that it nevertheless moved up the trunk like normal sap. After three days he replaced the picric acid with a red dye (fuchsin). This moved slowly up the stem, staining as it went. Evidently its passage was not impeded by any cells which had been killed by the acid.

Simple ringing of the stem, cutting out the phloem, does not interrupt the upward flow of water. If the xylem of cut shoots is sealed off with wax the plant wilts rapidly, whereas the sealing of the phloem has little effect on the turgidity of the shoot cells.

Translocation of dissolved minerals. The ascending xylem sap carries with it many mineral salts most of which are utilised in the metabolic activities of the leaves and shoot apices. Translocation is strikingly demonstrated by the use of radioactive tracers. Fig. 6.1 shows how radioactive phosphorus P^{32} is moved rapidly from an external solution up into the young leaves in balsam cuttings. That the tracer moves along in the xylem stream is easily shown by ringing experiments. Fig. 6.2 shows the results of one such experiment on young rooted

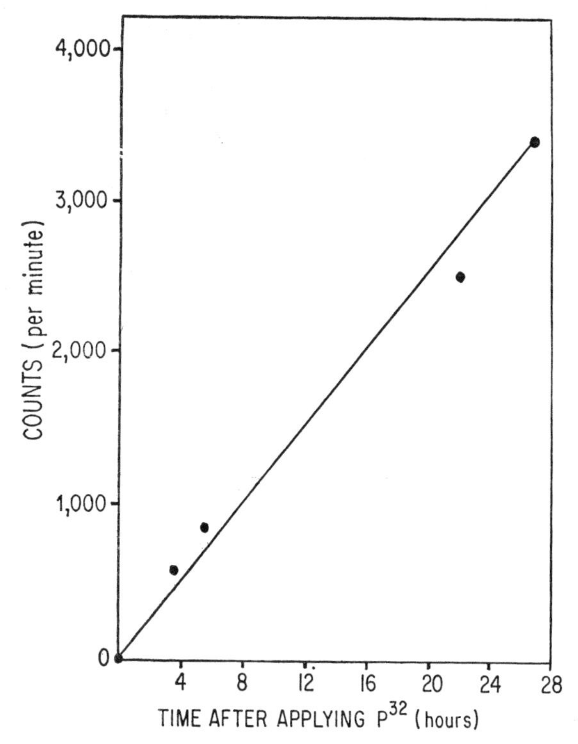

FIG. 6.1. Uptake of radioactive phosphorus (P^{32}) into the young leaves by a rooted cutting of balsam (*Impatiens sultani*)

cuttings of *Skimmia japonica*. There is little difference in the uptake by treated and untreated plants.

A critical experiment to determine the relative importance of the xylem and phloem in translocation of mineral salts was carried out on willow shoots by Stout and Hoagland in 1939.

One shoot was treated as follows and another was left untreated, as a control. The bark with the phloem was carefully separated from the wood by cutting longitudinal strips, leaving the ends of the strips still attached. A layer of paraffin-waxed paper was then inserted around the stem between the outer

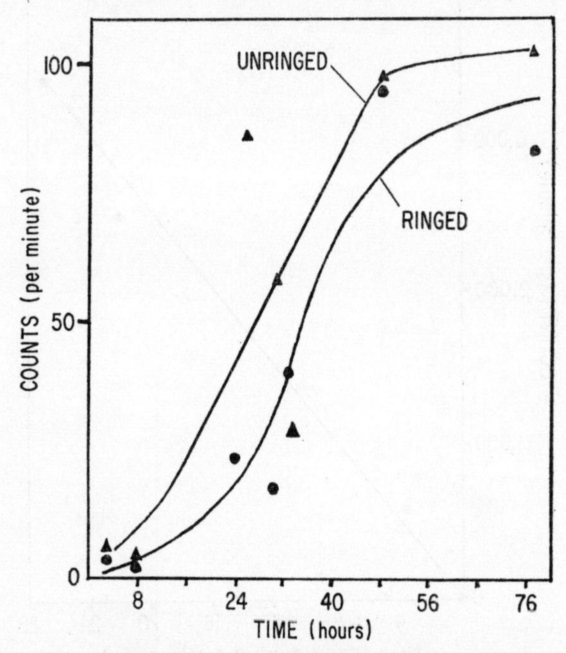

FIG. 6.2. Uptake of radioactive phosphorus (P^{32}) into the young leaves by ringed and unringed cuttings of *Skimmia*

layers and the wood. The whole was then sealed together (see Fig. 6.3).

The shoots were placed in a solution containing radioactive potassium (K^{42}) for a few hours. The parts of both shoots were then analysed for their K^{42} content; the results are given in Table 6.1.

TABLE 6.1

The distribution of potassium in the stem of willow (parts per million) after an absorption period of five hours

Position	*Treated branch*		*Intact branch*	
	ppm in outside	*ppm in wood*	*ppm in outside*	*ppm in wood*
Above strip	53	47	64	56
Stripped section	0·7	112	87	69
Below strip	84	58	74	67

They found, surprisingly, that the phloem of the control shoot which was intact always contained more tracer than the xylem. Yet when the two were separated the phloem contained very

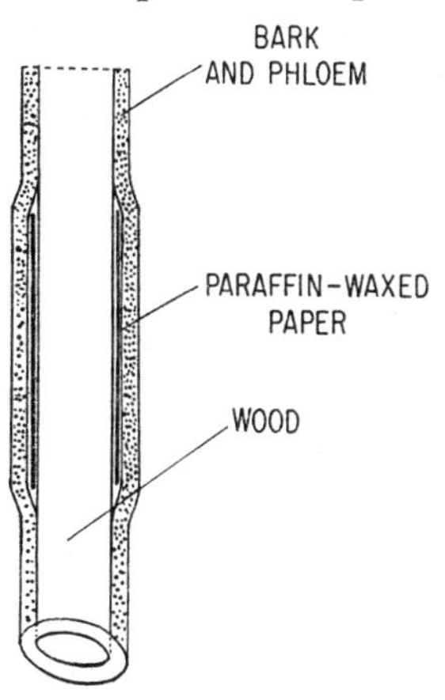

FIG. 6.3. Diagram to show the separation of living tissues from wood in willow

little. The conclusion must be that the phloem accumulates the potassium ions which the neighbouring xylem translocates to it.

Translocation of dissolved organic materials. Experiments on the transport of carbohydrates using C^{14} as the tracer prove that the xylem is of little significance in the translocation of dissolved organic substances, for these living tissues of the phloem are the principal transporting tissues. This may involve movement of the colloidal cytoplasm, with substances dissolved in the water–disperse phase of the system, as well as diffusion and various other metabolic processes.

Movement of water through the plant may be a bulk movement in response to transpiration loss, or it may be the less clearly defined translocation of dissolved substances. There are many puzzling aspects about these systems, one of which is the extraordinary speed at which both water and solutes move.

The apoplast and the symplast

In attempting to understand some of these problems, it is first necessary to have an understanding of some anatomical aspects of roots, stems, and leaves. It would be out of place to describe the salient features of the anatomy of these structures here. What is more to the point is to note those areas where water can move freely and their relation either to impervious areas or to those containing cytoplasm. In a recent review (1963) Salisbury draws a clear distinction between the two areas, calling those where free movement of water is possible the *apoplast* and those where the movement is not free and through the cytoplasmic system, the *symplast*. Figs. 6.4, 6.5, and 6.6 show the distribution of these areas in part of a root.

The apoplast corresponds to the *free-space* of any tissue and offers no metabolically controlled barriers to water movement. The cell wall, the intercellular spaces and the xylem system are all part of the apoplast. A nearly continuous water system may run from the cells of the outer cortex of the root, perhaps even from the soil solution, right through the plant to the cells of the leaf which are losing water by evaporation.

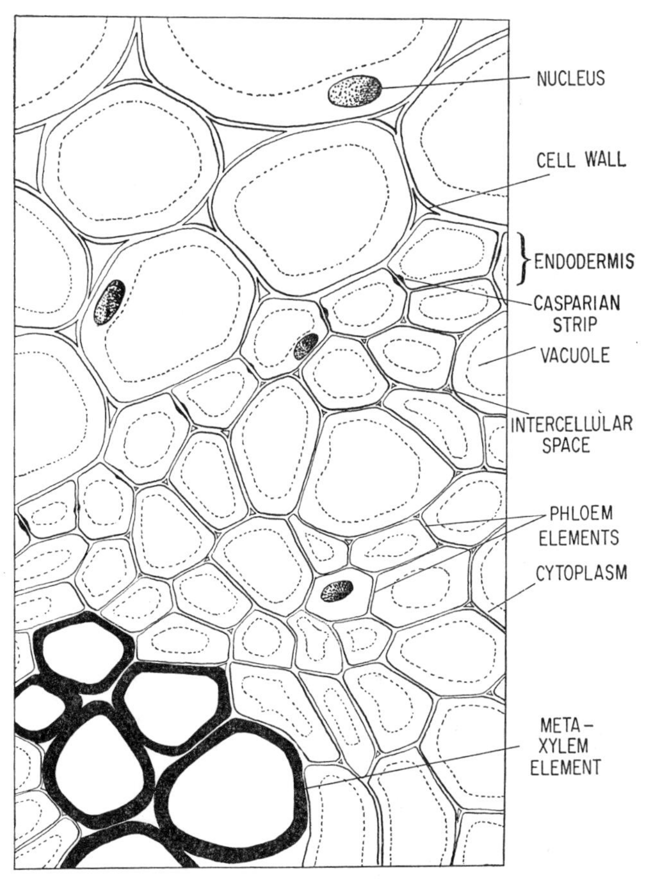

FIG. 6.4. Part of a transverse section through a young root of the buttercup (*Ranunculus*) (highly magnified)

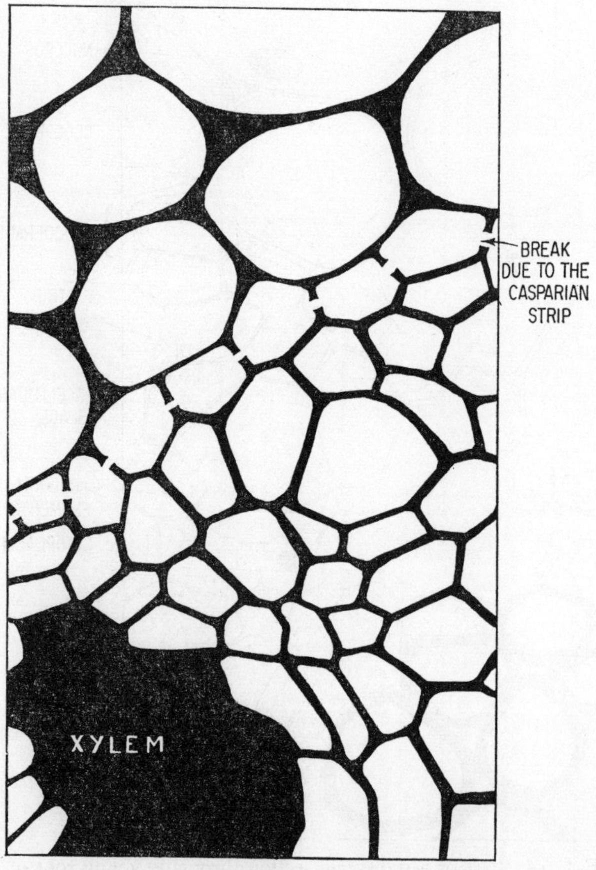

FIG. 6.5. The apoplast of the buttercup root

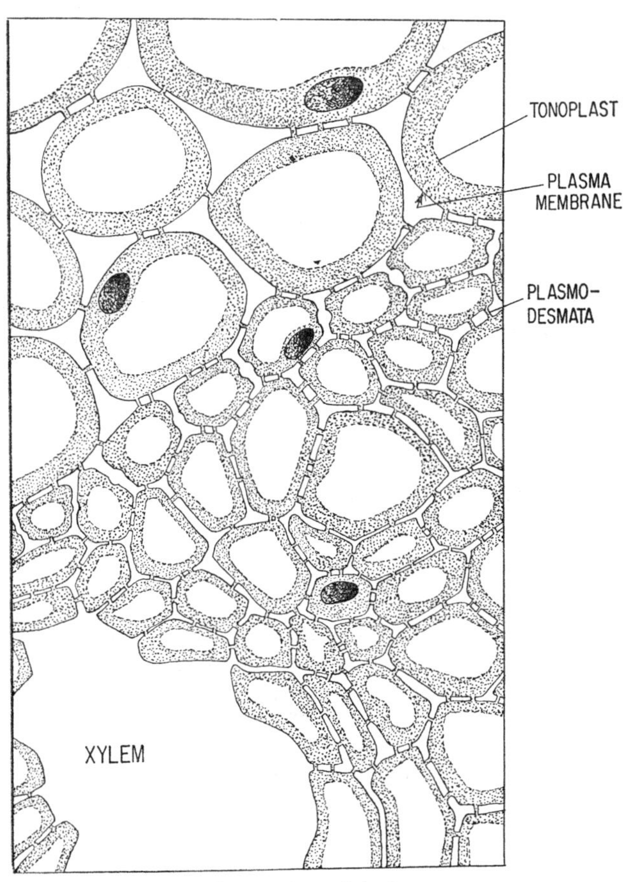

FIG. 6.6. The symplast of the buttercup root

It is possible that the apoplast is continuous from undifferentiated cells near the root apex to the young xylem elements. The only place where this system may not be continuous is across the endodermis. Most endodermal cell walls are impregnated with a lignin thickening, known as the *Casparian strip*, which is made partially impermeable by the deposition of suberin.

The symplast is the cytoplasmic system and is continuous through a great deal of the plant due to small cytoplasmic connections through the cell wall known as *plasmodesmata.* These are particularly well developed in the cytoplasmic connections that pass through the sieve plates of the phloem sieve tubes. The symplast includes no dead cells such as those of the xylem.

The energy requirements for the transport of solutes

Strasburger was the first to demonstrate the speed at which substances move through the plant. Table 6.2 gives some recent data for a variety of substances.

TABLE 6.2

The rate of movement of a variety of substances in various plants

Type of plant	*Substance*	*Pathway*	*Rate* (cm per hr)
Mesophytic deciduous tree	Water	Xylem	between 50 and 4360
Velvet plant (*Gynura aurantiaca*)	Sucrose	Phloem	between 20 and 100
Cocklebur (*Xanthium pennsylvanicum*)	Flowering hormone	Phloem (?)	between 0·25 and 0·42

There is considerable variation in the rates of movement of different substances through the same plant. Diffusion alone is insufficient to account for these speeds. There are three possible sources for the energy requirements of these movements: hydrostatic pressure differences, osmotic potential gradients, and energy made available by metabolic activities.

Movement of water in the transpiration stream is caused largely by hydrostatic pressure differences. There is a reduction in pressure around the leaf mesophyll cells due to evaporation. The pressure in the root cells is greater than that in the mesophyll cells, and so a mass flow of water takes place from root to leaf.

Rapid transpiration also results in a high water deficit in the cells of the leaves as the vacuoles and probably, to some extent, the cytoplasm lose water. At the end of the day, when transpiration is reduced, the high osmotic potential of these leaf cells will enable water uptake to continue for quite a time. Temporary conditions of this kind are common in many parts of the plant.

Evaporation is not the only cause of osmotic potential changes. Metabolic activity may result in the production of solutes which diffuse into the vacuole, causing a rise in the osmotic potential of the cell. Alternatively the osmotic potential may be lowered due to utilisation of vacuolar solutes.

Solutes may be carried along quite fortuitously with the water passing through the apoplast. They may also be conveyed in the symplast system by much more complex systems which, in the long run, are usually dependent on the production of energy in the form of adenosine triphosphate (ATP), within the cell.

Movement of dissolved organic materials

Partly because of the curious structure of the phloem elements, phloem transport is a much more difficult problem than that taking place in the xylem. The phloem elements are made up of sieve tubes and companion cells; the former are most directly concerned with translocation. The sieve tubes are devoid of nuclei and have their end walls, the sieve plates, pierced by numerous plasmodesmata. A normal sieve tube is only about 70–100 μ long. The phloem tissue thus presents a great many barriers to solute movement. Phloem sieve tubes are very susceptible to damage and even when undamaged may be blocked by the deposition of the carbohydrate *callose* around the lining of the pores in the sieve plate.

Münch's hypothesis

In 1930 the German physiologist Münch proposed a pressure-flow theory to account for the energy sources for solute movement. He suggested that when a leaf cell produces osmotically active substances, e.g. sucrose and glucose, their synthesis increases the osmotic potential of the cell so that water is taken up from the surrounding apoplast. This increases the pressure inside the cell and causes a mass flow of material from the producing cell in the leaf. This mass flow is continuous through the symplast system via the phloem of the vascular system (see Fig. 6.7), and can feed consumer cells either below the original producer cell in the root or stem, or above it, in the stem meristem, flowers, or fruits.

On reaching the consumer cell, the mass flow will cause an increase in cell size which will in turn affect the adjacent apoplast system. The increase in pressure here will favour an upward movement through the xylem, which will also be brought about by a reduction in pressure in the apoplast system surrounding the original producer cell. The continuous production of sugars by the producer cell and the continuous utilisation of solutes by the consumer cell provides the means whereby the cyclic system is maintained.

Münch was able to demonstrate his ideas with the simple apparatus shown in Fig. 6.8. The osmometers represent the producer and consumer cells and are connected as shown. The outer solutions represent the apoplast. The solid dots represent sugar molecules; the open dots represent various inorganic ions carried along in the mass flow, some of which may be able to pass through the differentially permeable

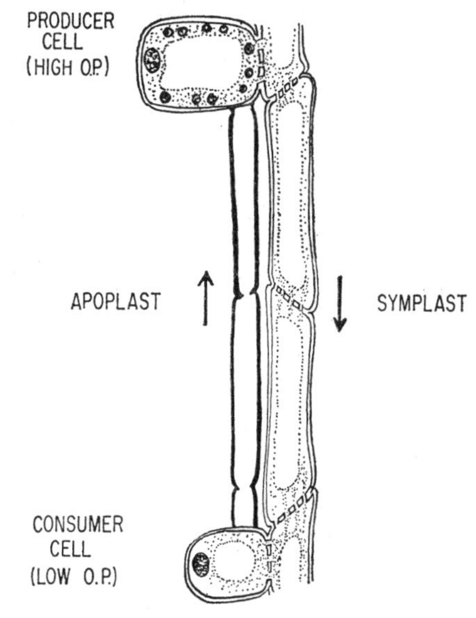

FIG. 6.7. **Diagram showing** the essential cells of **Münch's hypothesis**

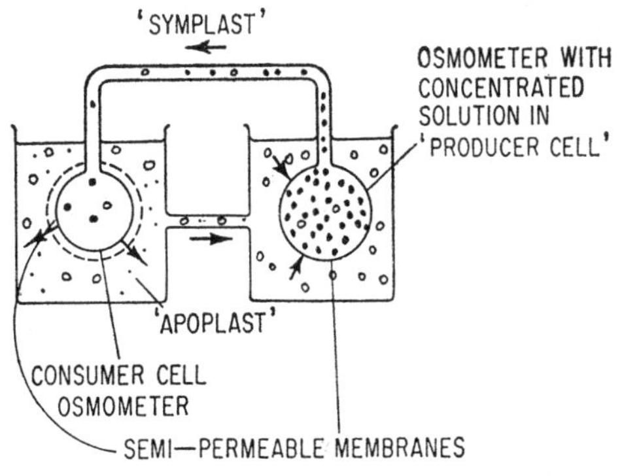

FIG. 6.8. **Apparatus used to demonstrate** Münch's hypothesis

membranes of the osmometers. Water, taken up by the osmometer containing the stronger solution, will move through to the second osmometer. As the pressure in the second osmometer increases it will cause it to swell and water will also diffuse out. These two effects will cause an opposite flow in the outer part of the system. The model will of course reach equilibrium when the distribution of solutes in the osmometers becomes equal, a situation which is unlikely to occur in the living plant.

Testing Münch's hypothesis

To test Münch's hypothesis we need to demonstrate an *osmotic gradient* from producer to consumer cell. Zimmermann in 1957 made incisions in the bark of a large ash tree and collected the phloem contents. A variety of sugars was found and the results of his experiment are indicated in Fig. 6.9. These readings showed a very satisfactory osmotic gradient in the summer when the trees are in leaf, though after leaf fall the gradient was much less marked.

The phloem contents should also be under a *positive pressure*. Measurements of pressures and pressure gradients are not particularly easy as the phloem is so susceptible to damage. Recently, Mittler in the U.S.A. employed aphids (*Longistigma caryae*). These were allowed to pierce the phloem with their stylets and provided a source of phloem exudates obtained with a minimum of damage to the tissue. If the insect is left in place it feeds off the phloem liquids and excretes surplus in the form of *honeydew* from time to time. Mittler removed the aphid but left its stylet in position which continued to exude sap, often for several days. This showed that the phloem contents were under pressure (unlike the xylem elements which are normally under tension). The quantities of exudate were extraordinarily high: a single stylet could exude 5 cu. mm per hour. This would give a translocation velocity of the order of 100–150 cm per hr. These figures indicate that the contents of *a single sieve tube* must move through the sieve plate *in less than a second*; a truly remarkable finding. Calculation of the area of the sieve pores shows that they are only about 1/11 of the area of the whole sieve plate. This works out at a velocity of movement through the pores of about 16·5 m per hr!

Finally, if Münch's hypothesis is to hold good, there should be a mass flow of material through the phloem which could

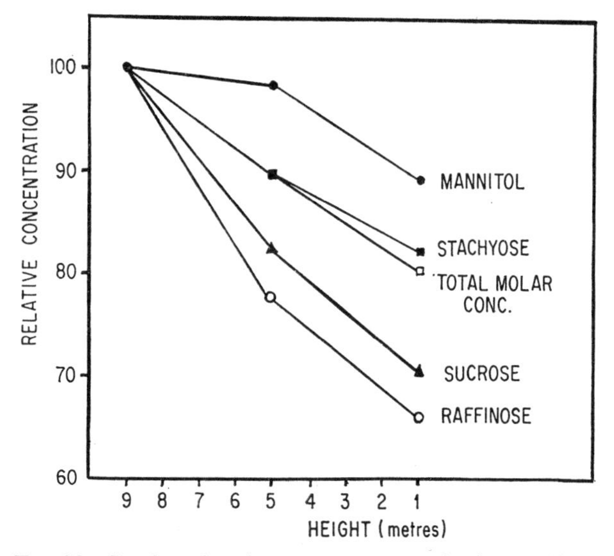

FIG. 6.9. Graph to show how sugar concentration in the phloem exudate of the white ash (*Fraxinus americana*) varies with sample height

only go in one direction, though it could be up or down depending on the site of the consumer cells. If there is a mass flow then the various solutes should move together, in the same direction and at much the same rate, though there is no reason why some should not be metabolised on the way. Recently Biddulph and Cory in the U.S.A. investigated this movement aspect with radioactive tracers. They labelled the sugars being formed in the leaf by allowing photosynthesis to take place in an atmosphere containing carbon dioxide labelled with C^{14}. Tracer phosphate (P^{32}) was applied to the leaf situated below that which was treated with the labelled carbon dioxide. These two tracers can be distinguished because of the different penetrating powers of their radiation; phosphorus is more penetrating than carbon. In addition, carbon has a much longer half-life (5500 years) than phosphorus (14 days) and so the phosphorus can be allowed to decay away and the residual carbon identified in the usual way, by autoradiography.

At first sight their results seemed to disagree with the hypothesis as they showed that bidirectional movement often occurred. However, careful analysis of the autoradiographs showed that movement usually took place in only one direction in an individual vascular bundle. More rarely bidirectional movement took place within the same phloem element. In the latter case Münch's hypothesis can no longer be tenable. Biddulph and Cory then notice that such bidirectional movement only occurred in young phloem elements. This was particularly interesting as these are the only phloem elements in which *protoplasmic streaming* or *cyclosis* has been observed. It seems that Münch's pressure–flow ideas hold good for mature phloem elements, but the more complex streaming systems may allow for the bidirectional translocation in the young phloem cells of meristems and vein endings.

Further problems

We have seen that the rapid transport of dissolved organic substances through the phloem can be accounted for by a pressure-flow system. We also know that the phloem must be alive if translocation is to take place. It has often been suggested that the rapid flow of material through the sieve plates could be aided by increased metabolic activity. Temperature, which affects respiratory rates and energy release, might also affect the rate of translocation. Surprisingly, investigations into the effect of temperature have produced conflicting results. Reliable data shows that increased temperature does indeed, in some cases, speed up the rate of movement. Equally reliable evidence shows that sometimes increased temperature has the opposite effect. There seems no obvious reason for this anomalous fact.

Crafts has put forward a suggestion that the rapidity of phloem translocation was facilitated by water molecules in the sieve plate existing in a state of *superfluidity* with their molecules ordered in some way so as to reduce viscosity. If such a system does operate, we can only find out how by studying three problems: first, how the water molecules could be arranged so as to produce such a superfluid state; second, whether energy would be needed to create such a condition; and third, whether examination of the ultra-structure of the sieve plate and cytoplasm of the sieve elements can give us any further clue.

At the moment then, beyond the fact that Münch's pressure-flow system probably operates, there are a large number of enigmatic aspects concerning the movement of dissolved organic materials in the plant.

7

The Paths of Water Movement

Movement of water in the leaf

For any detailed discussion of the movement of water and dissolved minerals through the plant, it is best to begin at the site of evaporation where most of the energy needed for water absorption and transport is made available. Fig. 7.1 shows a diagram of a longitudinal vertical section through the lower part of the leaf of the garden privet (*Ligustrum ovalifolium*), which gives an indication of the cell structure concerned.

It used to be thought that the passage of water took place from cell to cell by a system of osmotic gradients (route *abcd* in Fig. 7.1). The cell that was losing most water (perhaps *a*) would have the highest osmotic potential and this would result in a flow of water from the neighbouring cell (*b*). This would, in turn, gain water from another cell of still lower osmotic potential (*c*). Finally, the lowering of hydrostatic pressure around the fine xylem endings (*d*) would result in a flow of water up the stem and petiole.

Alternatively, there seems no obvious reason why water loss may not also be replaced by mass flow via the intercellular spaces and fully permeable cell walls of the apoplast system (route *a′b′d*). Which pathway is in fact used must depend at least to some extent on the rate of transpiration and the availability of water. If plenty is available, it is probable that the apoplast route is the primary one. On the other hand, if there is an insufficient supply in the xylem, then water will be taken from the vacuoles and osmotic forces will be the crucial ones determining water movement.

In 1963 Professor Weatherley of Aberdeen carried out some

interesting experiments to investigate the plausibility of these alternative pathways; the following details are quoted with his permission: If water passes through the cell contents, the numerous cytoplasmic membranes which have to be crossed

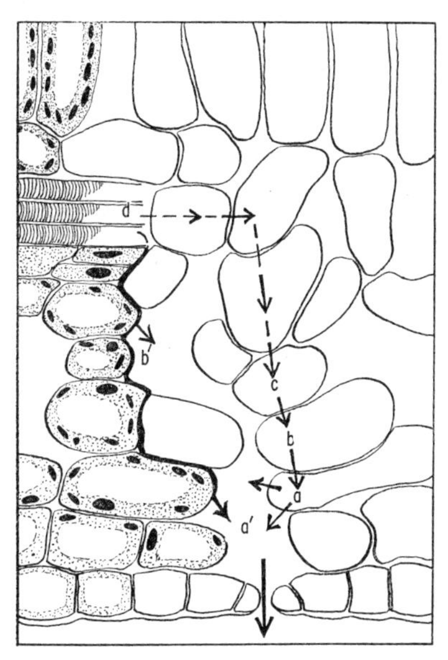

FIG. 7.1. Diagram of the lower part of the privet leaf seen in vertical section

will act as resistances to movement. In addition, any factors affecting the state of the cytoplasm must similarly affect the movement of water. Professor Weatherley has constructed two hypothetical models of leaf cells and also tested his hypotheses by experiments on the living leaf.

Fig. 7.2 is a model of a transpiring cell (as *a* in Fig. 7.1) on the hypothesis that the water moves *through* the vacuoles. The apparatus is filled with water. V represents the vacuole and the capillary tube, *c*, the resistance presented by the transpiration stream lying below the cell. The bulb *b* is porous. Evaporation from it is replaced by water drawn from the reservoir *w* through the tube *c*. This tube imposes a resistance to flow causing a reduction in pressure in V so that mercury is drawn up the vertical tube. This reduction of pressure represents the depression of water potential in the cell. For a given rate of transpiration a certain reduction of pressure will be manifest

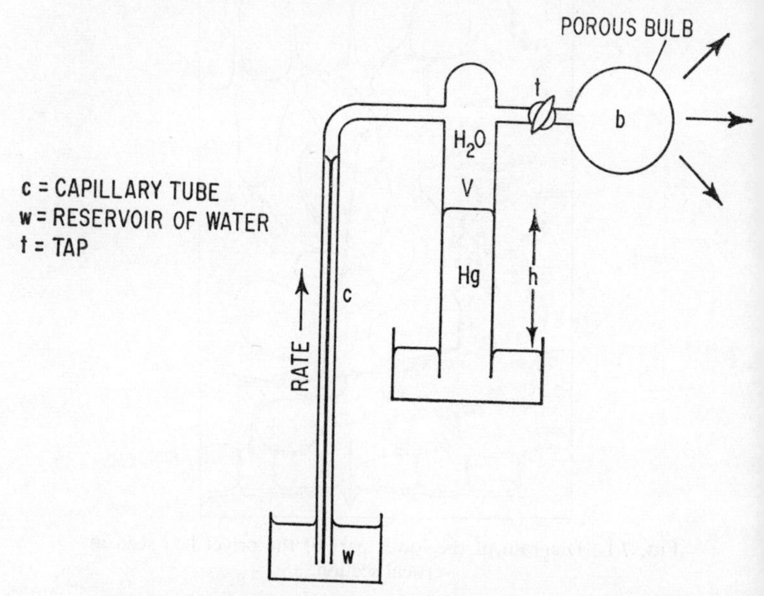

FIG. 7.2. Hypothetical model of transpiring cell *a* in Fig. 7.1 if water moves through the vacuoles

by the mercury reaching a steady height *h*. The volume of the mercury occupying the length *h* in the vertical tube can then be calculated and gives an indication of the water deficit of the cell.

If the tap *t* is now closed, i.e. evaporation stopped suddenly, water will continue to move into the cell through *c* whilst the mercury in the vertical tube falls. When *h* finally drops to zero, uptake through *c* will cease and the cell will have reached saturation point. During this fall in uptake the rate of uptake f_t at any instant will be proportional to h_t at that instant,

$$f_t = \frac{h_t}{r} \quad \text{(i)}$$

where r is a constant proportional to the resistance of the tube c. Hence the rate of uptake f_t plotted against the height of mercury h_t will give a straight line as in Fig. 7.3. It can be

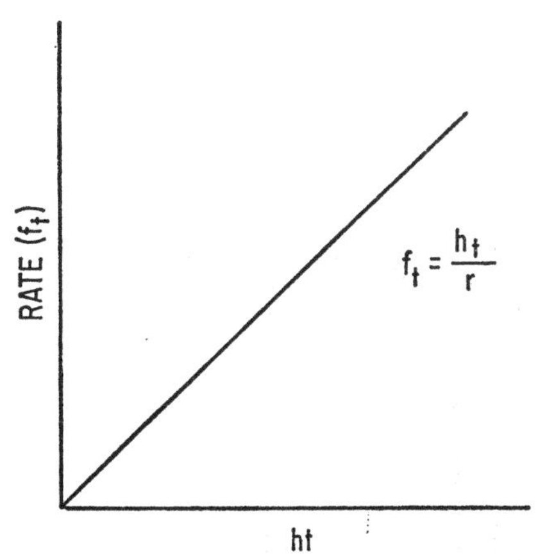

FIG. 7.3. Graph to show how rate of water uptake is related to the height of mercury in the model

shown that the die-away in uptake is logarithmic and may be represented by the equation:

$$f_t = f_0 e^{-t/r'} \quad \text{(ii)}$$

where f_0 is the steady rate of flow (evaporation) up to the instant of turning off the tap and r' is a constant similar to r. Thus a plot of f_t against time will give a logarithmic decay curve as in Fig. 7.4, and plotting the logarithms of f_t against time will give a straight line (Fig. 7.5).

Weatherley's alternative hypothesis (route $a'b'd$ in Fig. 7.1) assumes that water passes round the outsides of the cells, that is, through their cell walls and inter-cellular spaces of the apoplast system. The implication is that there is a resistance

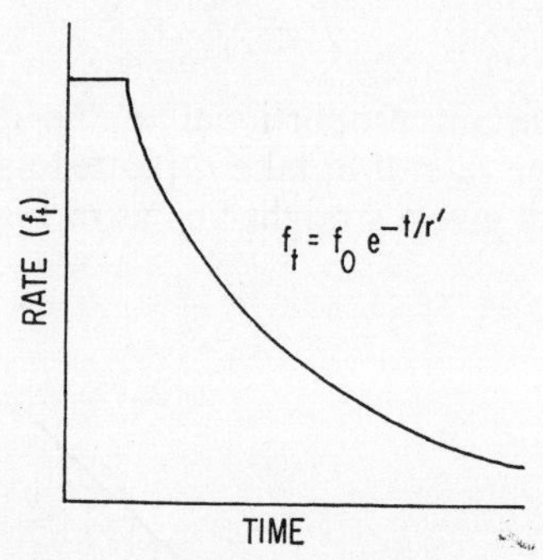

FIG. 7.4. Graph to show that the die-away of water uptake follows a logarithmic curve when the tap *t* is closed

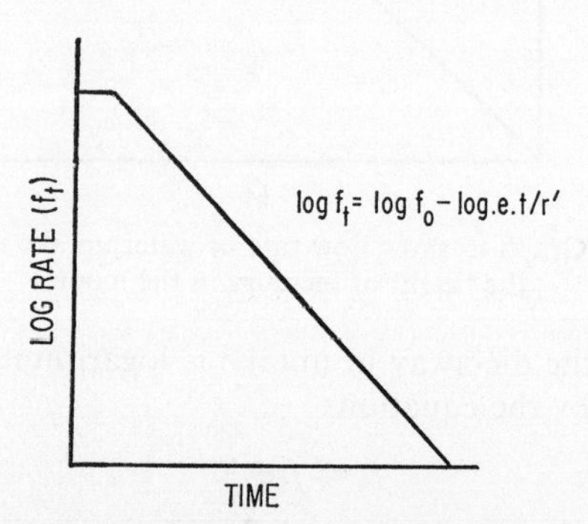

FIG. 7.5. Graph to show the relationship between log rate of uptake and time

between the pathway and the vacuole. Such an additional resistance is shown in Fig. 7.6 in which the vacuole v is separated from the main stream by a capillary c' of resistance

r''. With steady evaporation the mercury will rise as before and attain the same steady position in relation to f_0 and r as it will still measure the reduction of pressure h at the top of c'. Closing the tap will, however, have an entirely different effect. During the period of steady transpiration water is dragged

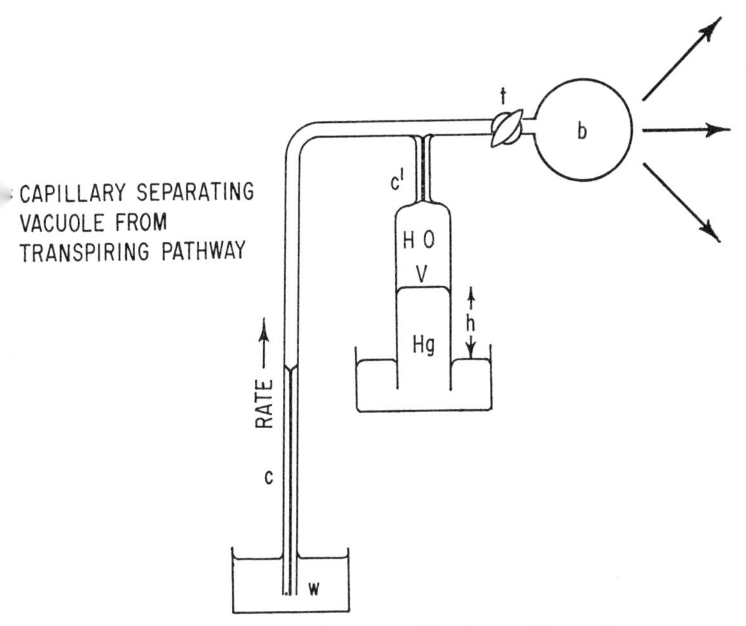

FIG. 7.6. Alternative hypothetical model of transpiring cell *a* or *a'* in Fig. 7.1

through the capillary tube c alone, whereas after the tap has been closed and the mercury begins to fall, water must be dragged through both capillary tube c and c'. Thus rate of uptake will instantaneously fall from that given by equation (i) to:

$$f' = \frac{h}{r + r''} \qquad \text{(iii)}$$

After the instantaneous fall, f will decline logarithmically according to equation (ii). Thus the curve of rate of uptake f_t

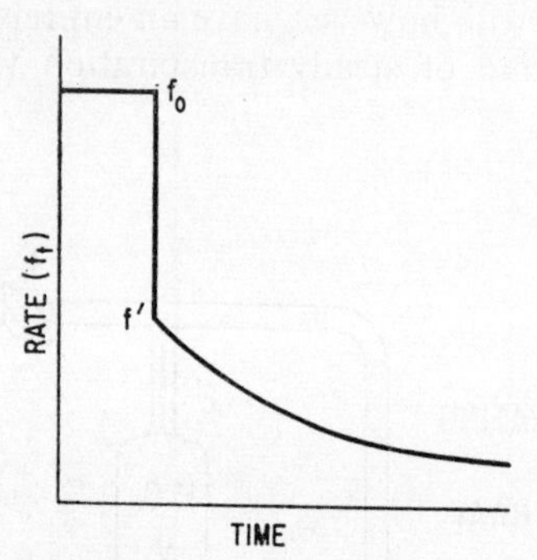

FIG. 7.7. Graph to show that the die-away of uptake follows a logarithmic curve when the tap t is closed

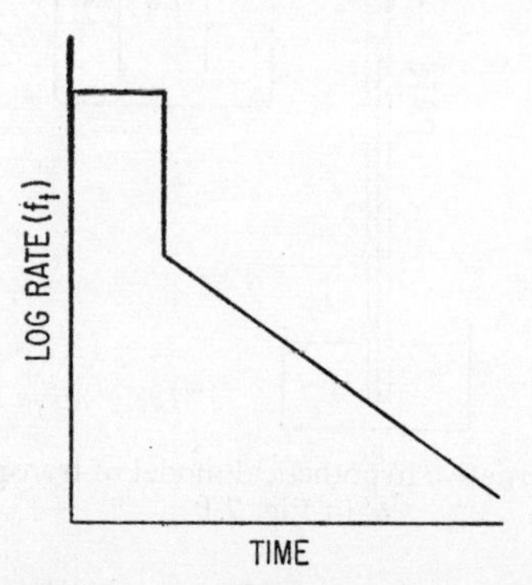

FIG. 7.8. Graph to show the relations between log rate of uptake and time

plotted against time will appear as in Fig. 7.7. The logarithmic plot appears as in Fig. 7.8.

Weatherley set out to check these hypotheses by observations on the living plant. Data was obtained using leaves of geranium (*Pelargonium zonale*), *Populus gileadensis* and flowering currant (*Ribes sanguineum*). In each experiment a detached leaf was fitted into a simple potometer as shown in

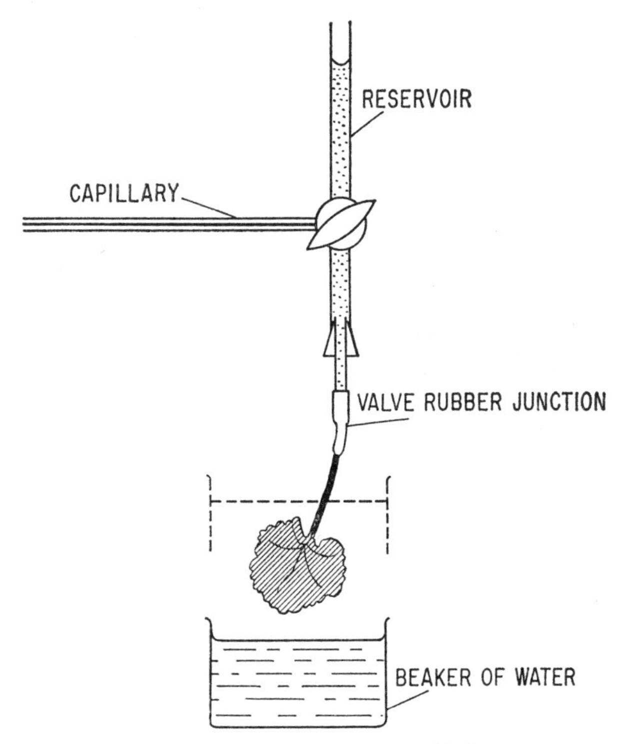

FIG. 7.9. Weatherley's simple potometer with its arrangement for stopping transpiration by immersion

Fig. 7.9. Uptake was measured by following the movement of a meniscus in the horizontal tube using a travelling Vernier microscope. For one or two hours after measurements began a steady measured rate of transpiration was maintained. Transpiration was then abruptly halted by immersing the leaf in water or medicinal paraffin. In the course of each experiment the total amount of water needed to saturate the leaf was measured by the potometer. This quantity of water represented the initial water deficit of the leaf. Further, the water

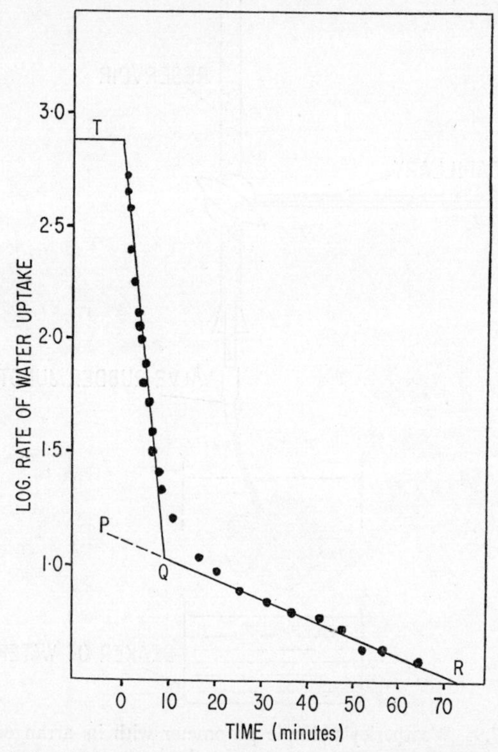

FIG. 7.10. Die-away curve of the rate of uptake by a leaf of geranium (*Pelargonium zonale*) on stopping transpiration

deficit at any time was measurable as the amount of water subsequently absorbed to attain full saturation. A typical result is shown in Fig. 7.10. Clearly this result fits the second hypothesis, that is, that water moves around the outsides of the cells rather than through them.

ATE 5. Californian redwoods (*Sequoia sempervirens*) in Muir Woods, San Francisco, California. These trees are often over 300 ft high.

PLATE 6. Leaves of the giant water lily (*Victoria amazonica*) with the water hyacinth (*Eichhorni*
Courtesy: Amateur Gardening

The experimental results do however differ from those predicted by the second hypothesis. The drop in rate of water loss is not instantaneous if transpiration is stopped suddenly. This seems to be due to a water deficit developing in the pathway itself. This uptake by the pathway appears to decay logarithmically, like that of the inner space of the cells, but much more rapidly, indicating a low resistance between itself and the water supplied at the cut end of the petiole. On this interpretation, the initial steep fall in Fig. 7.10 represents the decay in uptake by the apoplast and the subsequent slower fall represents the decay in uptake by the inner space (vacuoles and probably cytoplasm).

If we assume that the pattern of uptake by the inner space of the cells was uniform throughout, its decay curve during the first ten minutes will be represented by the line of extrapolation to P. The line PR would then represent the decay in uptake by the inner space throughout the experiment, and the line PQ could be taken as a representation of the rates of uptake by the inner space alone during the first twenty minutes of the experiment. By subtracting these values from the total uptake Professor Weatherley obtained the rates of uptake by the apoplast alone. These are shown in Fig. 7.11. It will be seen that whilst the apoplast becomes virtually saturated after twenty minutes, the inner space would have taken several hours to reach a similar state. Evidently detached leaves show relatively slow changes in the symplast and inner space when rapid fluctuations occur in water content of the pathway. From the point of view of the present discussion it is more significant that the rate of uptake into the inner space, represented by P, is only 1/60 of the transpiration rate (in *Pelargonium zonale*), that is to say, *by far the greatest proportion of water in the leaf moves through the apoplast.*

Mass flow through the cell walls would be expected to be less sensitive to temperature, while that through the cytoplasmic membranes, whose permeability is often much reduced by a lowering of temperature, might be expected to show marked effects. To test this, Weatherley stopped leaf transpiration suddenly as before by immersing the leaf in water at about 3°C. The results of such an experiment are shown in Fig. 7.12 in which logarithms of rate of uptake are plotted against time. It will be seen that the rate fell away logarithmically with no

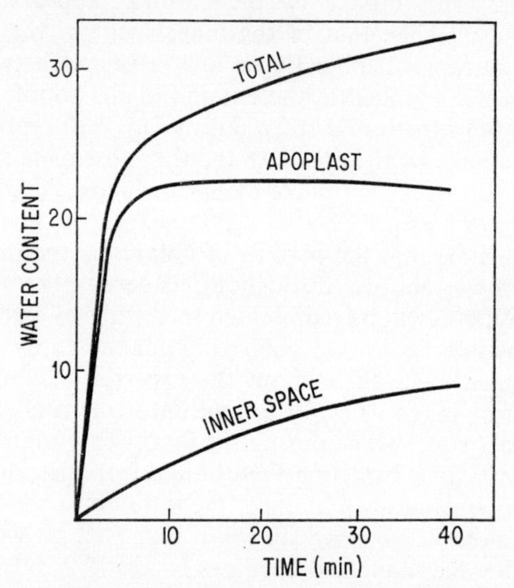

FIG. 7.11. Graph to show the progressive saturation of inner space and apoplast of the leaf of the flowering currant (*Ribes sanguineum*)

break until after 22 minutes when uptake had completely stopped. Thereafter no uptake was recorded for a period of 60 minutes. At the end of this period the cold water was replaced by water at 22·6°C. Uptake was resumed at once and after attaining a maximum, declined logarithmically as in the previous experiments. Uptake into the apoplast seemed to be little affected by temperature whereas uptake into the inner space was stopped completely at low temperature. This bears out the hypothesis that most of the water movement is through the apoplast.

Movement of water in the stem

Over two hundred years ago the English clergyman, Stephen Hales, in his *Vegetable Staticks*, a work that is usually regarded as the foundation of plant physiology, suggested that two mechanisms might be responsible together for the movement

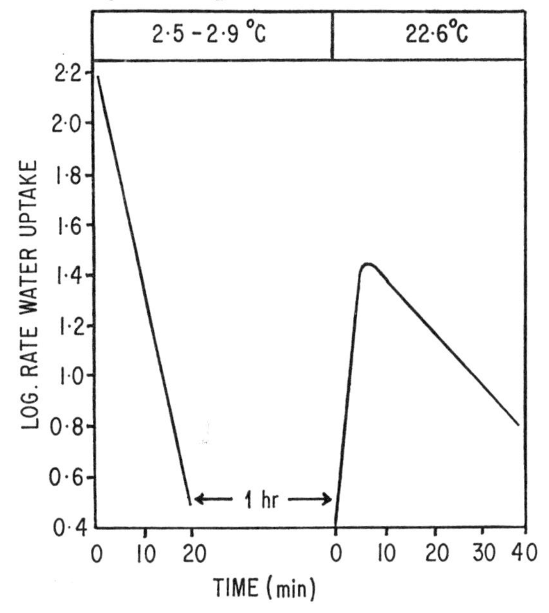

FIG. 7.12. The effect of temperature on water uptake into the leaf

of sap in stems and trunks; *root pressure* and *transpiration* or *shoot tension*. The contribution of root pressure is only moderately important; attachment of a manometer to a stem stump seldom produces a pressure even as high as two atmospheres. The rate of movement of water exuding under root pressure is also slow, though rates of over 100 ml per day have been recorded from the cut stumps of herbaceous plants (corn and sugar cane). Trees (paper birch) may produce as much as 28 litres per day. Root pressure seems to be at least partly under metabolic control. It may be caused by a release of solutes in the root system followed by an osmotic uptake of water; hence the exudation. Root pressure is of primary importance in guttation and in assisting bud break in spring, as both these conditions occur when transpiration is minimal. Root pressure could be taken as a background source of water movement which only comes into its own when Hales' 'shoot tension' is not operative.

Shoot tension. If most of the energy needed for the ascent of sap comes either from evaporation or from an increase in the osmotic potential of the leaf cells, we must accept the fact that a tension exists around the fine xylem endings in the leaf. How then is this transmitted to the root so that water can be drawn up? Suction pumps are incapable of drawing water up more than the height of the water barometer 10·35 m (34 ft) for at this height the pressure exerted by the column of liquid equals that exerted by the atmosphere.

Of course many trees are much taller than this. The tallest Douglas firs of the north west coast of America are sometimes over 92 m (300 ft) high, while in California the Wellingtonia (*Sequoia gigantea*) reaches more than 97 m (320 ft). Even bigger are the Californian redwoods (*Sequoia sempervirens*) (see Plate 5) one of which, in Humboldt County, is over 112 m (367 ft) high with a 15·2 m (50 ft) girth and is probably the largest tree in the world. Trees such as redwoods must be capable of drawing up water from well over this height as their roots may penetrate ten or twenty feet into the ground. The tallest trees in Britain are seldom more than 175 ft; these are usually of the same species as the tall trees of America.

In 1893 J. Böhm gave the clue to the problem by showing that fine columns of water can withstand considerable tensile stress. He set up a water-containing porous pot connected to a thin tube so that the tube dipped into mercury, as evaporation took place mercury was drawn up the tube to about 100 cm, yet the water column did not break.

As early as 1895 H. H. Dixon and J. Joly proposed their *cohesion theory* to account for the ascent of sap. They suggested that evaporation induced the tension and, due to the remarkable cohesive properties of thin columns of water, this was transmitted to the root. Recent investigations have shown that thin, air-tight tubes can withstand colossal tensions of 100 kg per sq. cm. Xylem sap contains a variety of dissolved substances, and though its cohesive properties are less than those of pure water in airtight capillaries, it appears to have sufficient tensile strength for the maintenance of very long continuous columns. In 1932 H. F. Thut in the U.S.A. demonstrated that a plant shoot was capable of producing a similar effect to that obtained by Böhm with his porous pot. Using a sprig of cedar carefully sealed into a glass tube he was able to get the mercury to rise 24·8 cm above the 76 cm mark. This is equivalent to a water rise of over 14 m.

Vessels and tracheids in the xylem system

In constructing models to demonstrate the cohesion–tension system it is easy to forget the scale of the cells of the vascular system. The xylem elements are probably much finer than can be constructed out of conventional materials. The Pteridophytes and conifers contain only the relatively short *tracheids* in their xylem. These are about 5 mm long and between 14 and 40 μ wide. Angiosperms may contain tracheids, but xylem is characterised by much longer *vessels*. These are often a metre long and between 300 and 500 μ in diameter. All these elements are thickened with lignin, which apart from giving rigidity to the whole stem, is vital in keeping the tubes from collapsing under tension. If an intact xylem element is carefully exposed and then pricked, the water columns running through it snap apart as the tension is released. In some cases the air entering the base of a cut stem can actually be heard to hiss.

Further evidence in favour of the cohesion–tension theory

Further evidence that the tension originates from the leafy

parts of the plant comes from the experiments of Huber in Germany in 1935. He investigated the movement of water in the xylem by an ingenious technique. He inserted a small heating wire into the wood; a few inches higher he introduced a small thermocouple. By taking the times for the warmed sap to be detected by the thermocouple, after switching on the heating coil for a few seconds, accurate measurements of the rate of flow could be obtained. This technique involved a

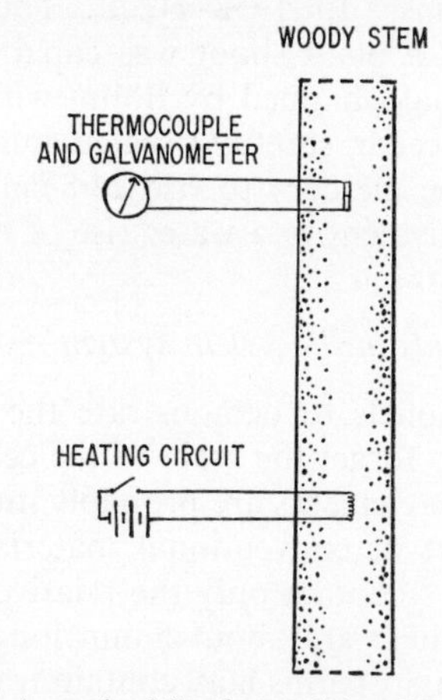

FIG. 7.13. Huber's method for determining the rate of movement of xylem sap

minimum of interference with the xylem elements. He found that the velocity of flow showed a diurnal variation which was similar to transpiration but that the rate of flow varied over different parts of the plant. Movement began in the young branches nearest the leaves, later it took place in the main trunk. Towards the end of the day it began to slow down just at the branches and later in the main trunk. Figs. 7.13 and 7.14 illustrate his method and results.

MacDougal in 1936 suggested that if the xylem is under tension while transpiration is taking place, then the diameter of the trunk should also show diurnal changes. He attached a device called a dendrograph to the trunk. This sensitive device measures any small changes in diameter. He found that the diameter of the trunk decreased high up the tree earlier in the morning. Later the diameter decreased lower down the tree. This result confirmed Huber's findings.

Of course, wind movements might cause breakage of the water columns. Frost can also have very serious consequences, because air is less soluble in ice. Freezing almost invariably

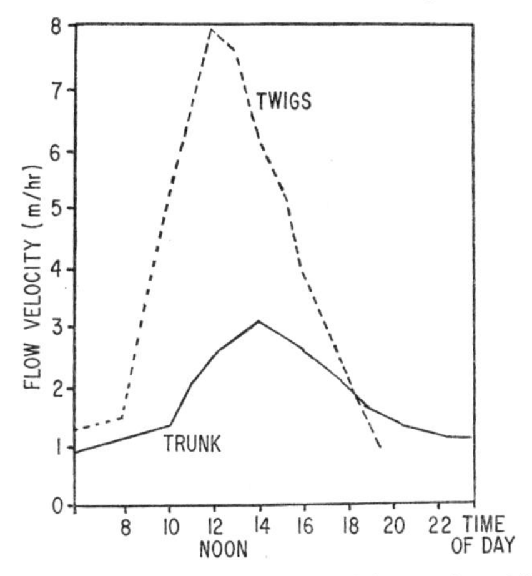

FIG. 7.14. Graph to show how time of day and position on the tree affect the rate of movement of xylem sap

causes air bubbles to appear in the xylem elements. This will occur when transpiration is low anyhow, and so the disruption of the xylem may not be serious. As the weather becomes warmer, the air may redissolve. Alternatively high positive pressures due to root pressure may cause the air to dissolve. In any case, as the tree begins growth, new xylem elements are laid down and these may by-pass the damaged pathways. This is particularly important in hardwoods (e.g. oak and ash) which have large vessels which are easily disrupted.

By taking freshly-cut lengths of trunk and forcing water through the wood, it is possible to gauge the force required to move the water through the whole trunk. The energy available due to evaporation can also be calculated. Table 7.1 gives typical values for the water potential of the various areas concerned.

TABLE 7.1

The water potential of various areas in and around the plant

Area	*Negative water potential or Diffusion pressure deficit* (atm)
Air at 90 per cent relative humidity, 20°C	140
Leaves	10–50
Roots	5–6
Soil Water	0·1

As it takes one atmosphere pressure to raise a column of water 34 ft and the resistance to water flow seldom exceeds twenty atmospheres for a tall tree, the figures in the above table show that there should be plenty of energy available for the upward movement of water.

Movement of water in the root

We have already discussed the problem of water uptake into the root. The account of the apoplast system has also indicated that free movement of water probably occurs through most parts of the young root unless it is prevented by the deposition of waterproofing materials such as cutin and suberin. The older root epidermis at about the level of the root hairs may have some of these materials deposited on the outside; still older roots possess a true corky layer which makes their outer layers completely impervious to water. The only other place of any importance where suberin is layed down is the Casparian strip of the endodermis which prevents passage of water through the intercellular spaces and cell walls. The endodermis is the one place where water must pass through the cytoplasm if it is to be drawn into the xylem endings. However, it is possible that a vertical movement through the cells near the growing apex could by-pass the endodermal barrier. There

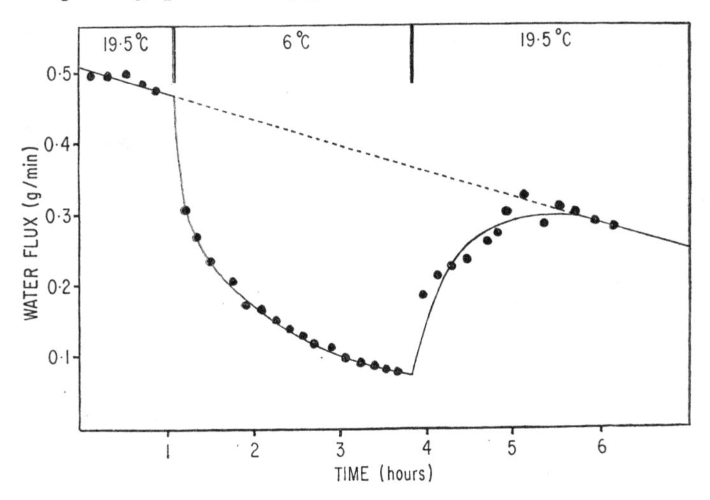

FIG. 7.15. Graph to show the effect of temperature on the movement of water through the root of tomato

are also unsuberised *passage cells* in the endodermis of many plants.

Professor Weatherley has investigated the factors affecting water movement in the root cortex by an interesting series of experiments. He based his experiments again on two alternate systems of mass flow of water. If mass flow occurs across the root cortex, i.e. if it follows the apoplastic, intercellular pathway, then the flux (movement) will be much more susceptible to pressure gradients than osmotic gradients. On the other hand, if there is a cytoplasmic barrier (e.g. at the endodermis) then osmotic gradients will have comparable effects to pressure gradients in the flux of water across the root cortex.

The experimental material was tomato plant stem stumps with attached roots, previously grown in water culture. They were placed in a pressure canister with their cut stems protruding. The apparatus was arranged so that the pressure around the roots could be varied by the use of compressed air. The liquid medium and with it the osmotic potential around the roots could also be varied. With a pressure gradient of 2 atm across the root, Weatherley was able to calculate that three-quarters of the water movement was due to osmosis and only a quarter was due to mass flow. He also found that the permeability of the root cells to the osmotic movement of water could change considerably under different conditions. The flux was very sensitive to temperature (Fig. 7.15). Use of cyanide, the respiration inhibitor, also upset the flow of water.

It seems almost certain that the site of this control lies in the root endodermis. The 25 per cent of water movement unaffected by cold or cyanide inhibition probably occurs through the passage cells, or unsuberised sections of the endodermis. The remaining 75 per cent of the water must pass through the cytoplasm of the endodermis under the influence of differing osmotic potentials. With this single interruption the flow of water from soil to air can be regarded as a continuous tensile stream. The cells surrounding the pathway adjust their water balance, often rather slowly, according to the various changing conditions.

8

Water and Plant Evolution

Water and the origin of life

It should be clear from the preceding chapters that life, at least as we know it, might be said to devolve on the water molecule. The various possibilities for the origin of life have been examined in detail by A. I. Oparin in 1957. He suggested that a number of stages had to be passed through for this process to have occurred. First the chemical substances necessary for life must have been formed. Second, they must have become organised into some self-reproducing structure or association of molecules. Third, they must have developed a feeding system. Fourth, they must have been able to obtain energy from their food so as to be able to reproduce themselves. Finally, they must have become organised into some sort of convenient building unit such as the cell, which is well-organised for growth and differentiation.

Formation of organic substances. Oparin suggested from spectroscopic evidence obtained from viewing other planets, that the atmosphere of the primitive Earth probably contained hydrogen, water vapour, ammonia, and methane. Nitrogen, oxygen, and carbon dioxide would have been extremely rare. S. L. Miller, working under the direction of Harold Urey (1953) suggested that many important organic substances could have been formed from these gases. Miller realised that these gases would be subjected to intense ultra-violet radiation and occasional electric discharge. These energy sources could well allow for the combination of the gases into organic molecules. Miller designed an apparatus, called a pyrosynthometer to test his hypothesis (see Fig. 8.1). A silent electric discharge was passed through the gases mentioned over a period of about a week. At the end of the period the liquid in the apparatus had become reddish and on analysis was found to contain a wide range of organic substances. These

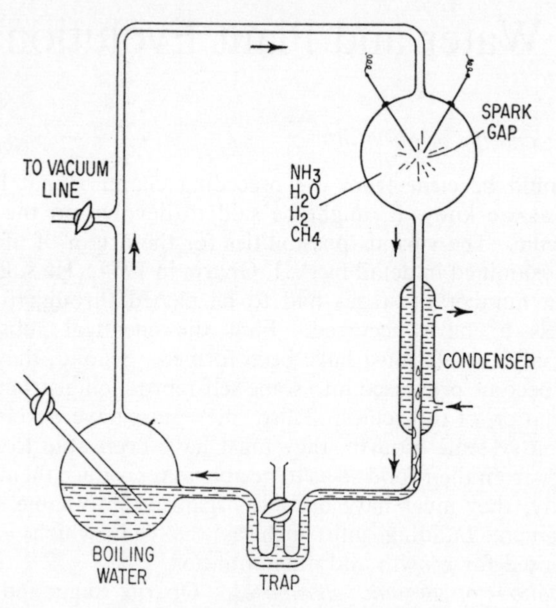

FIG. 8.1. Miller's pyrosynthometer

were principally amino acids (e.g. glycine, DL-alanine, sarcosine, and many others) also hydroxy acids (e.g. lactic, formic, acetic, glycolic, and propionic).

Formation of proteins. These amino acids could have been starting points for protein synthesis. In 1957 S. W. Fox showed that protein-like substances could be formed when a mixture of dry amino acids were heated together. This is a dehydration synthesis during which the water formed when the peptide bonds are made is volatilised:

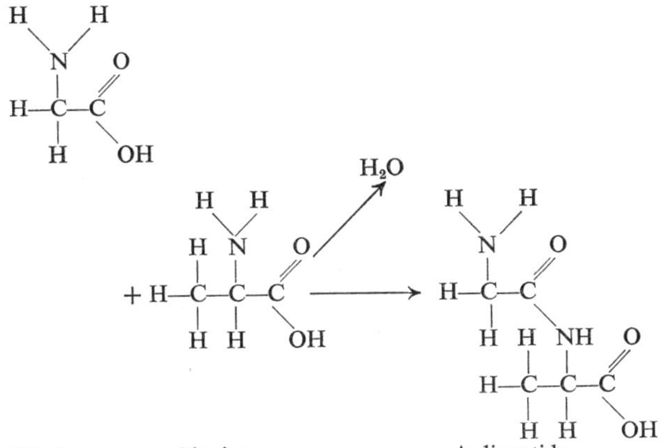

```
H   H
 \ /
  N   O
  |  //
H-C--C
  |   \
  H    OH
                      H2O
                      ↗
      H  H         /          H   H
       \/         /            \ /
     H  N   O   /               N   O
     |  |  //                   |  //
 + H-C--C--C    ———————→     H--C--C
     |  |   \                   |   \
     H  H    OH                 H  H NH   O
                                   |  |  //
                                H--C--C--C
                                   |  |   \
                                   H  H    OH
Glycine        Alanine            A dipeptide
```

This hypothesis for the pre-biotic formation of proteins seems to correspond to possible environmental conditions on the Earth some 800 million years ago, when it is thought that there was almost continuous rainfall bringing the various amino acids down onto the Earth. The Earth's surface was still hot, though, and between rainstorms, the drying out of the surface could have produced possible conditions for dehydration synthesis. Further rainfall and cooling would allow for the accumulation of these protein-like substances in the early seas to begin the formation of what J. B. S. Haldane referred to as a 'hot dilute soup'.

In addition to the formation of proteins, modern knowledge of cell biochemistry suggests that some molecules, such as the nucleic acids, capable of self-replication, would also have to be necessary for life to have originated. Some of the constituents of nucleic acids are easily synthesised. For instance, the base uracil found in ribonucleic acid, can be made by the combination of urea with malic and hydroxyacrylic acid.

Association of molecules. Accepting that such molecules could be synthesised in isolation, the problem remains of how they *associated themselves* in such a way as to form a series of organic molecules capable of growth and reproduction. The chances of such combinations occurring are more than rare. Nevertheless, given plenty of time, fruitful combinations could have occurred. Second, we tend to think of things decaying, either through the complex substances breaking down spontaneously or by oxidation, or by the action of micro-organisms. These two conditions were obviously less likely under primitive Earth conditions; oxidation could not occur because of the

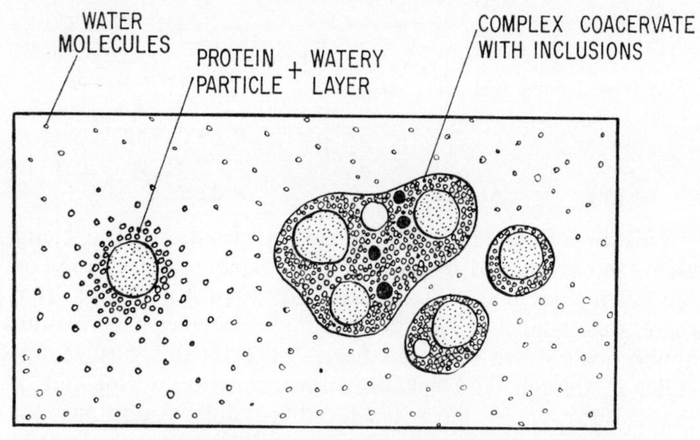

FIG. 8.2. Diagram to show the formation of a coacervate

absence of oxygen, and decay would not take place since there were no decaying organisms!

One of the answers to the problem comes from Oparin's observation that protein-like substances show 'spontaneous architectural tendencies'. When proteins are dissolved in water they become charged and attract water molecules (see Figs. 1.3 and 8.2). These associations tend to group together into more or less complex *coacervates* which Oparin believes provided the mechanism for the original concentration of organic substances. The simpler protein coacervates seem capable of trapping a wide range of other substances, as one can see using an aqueous mixture of the protein gelatin with the carbohydrate gum arabic under acid conditions. The coacervate formed appears extraordinarily amoeba-like.

Development of feeding systems. It is possible to visualise the better organised coacervates absorbing the lesser ones as a result of a chance collision. This initial feeding system would be essentially *heterotrophic*. If catalysts were present in the more developed coacervates these could break down the organic foods to release energy which they might use for syntheses of metabolic substances. The anaerobic breakdown

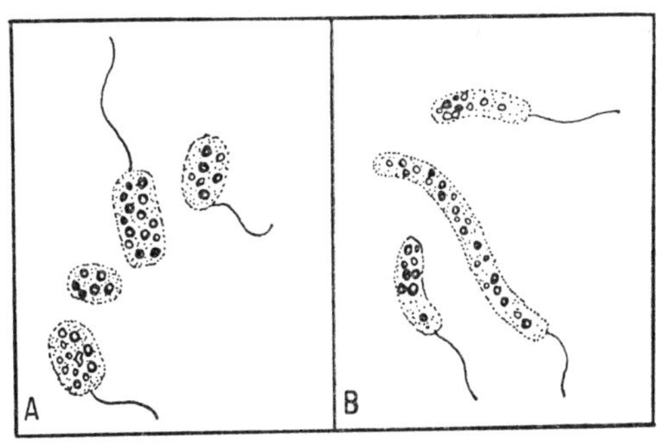

FIG. 8.3. Two photosynthetic bacteria. A: Chromatium. B: Thiospirillium (× 900)

of sugars would have been one way in which carbon dioxide might have accumulated in the atmosphere. This would have two important effects. First it would help form an insulating layer in the primitive atmosphere and second, it would form a source of carbon for a simple form of photosynthesis. The first autotrophes might well have been similar to present-day photosynthetic bacteria such as *Chromatium* and *Thiospirillum* (see Fig. 8.3) which are anaerobic organisms. These use hydrogen sulphide instead of water as their source of hydrogen, though whether this was the case in primitive organisms it is impossible to say.

$$H_2S + CO_2 \xrightarrow{light} CH_2O + S + \tfrac{1}{2}O_2$$

The oxygen produced in this process would eventually allow for aerobic respiration. The gradual accumulation of this gas in the atmosphere and its subsequent conversion to ozone in the upper layers would have provided an effective screen to the damaging ultra-violet rays which might have hampered further evolution on the Earth's surface. Some of the more highly adapted, aerobic and probably cellular organisms, of this era would probably have used a more advanced type of photosynthesis requiring water instead of hydrogen sulphide as a source of hydrogen. These organisms might well have resembled some of the simpler green algae found today.

Problems in the aquatic environment

As the thick cloud layers began to clear more efficient photosynthesis could take place and development of plant life may have progressed faster. Life in these early seas would have been quite different to that in the present-day sea. The concentration of dissolved substances must have been much lower and more like present-day fresh water, while the temperature may have been a good deal higher. Nevertheless the aquatic environment probably presented similar problems to the primitive living forms as it does to modern aquatic organisms.

The open sea

At the present time plant life in the open sea is almost exclusively planktonic; being composed of minute unicellular or colonial algae, most of which float in the upper few metres of the water. Larger species such as *Sargassum* may also be found, but these are fairly rare and have usually been broken off anchored specimens.

Any environment must physically support an organism. It must provide the organism with the gases, organic, and inorganic nutrients it requires. In many cases it must furthermore provide a situation suitable for sexual reproduction, and for the effective dispersal of the offspring.

In an aquatic environment water provides support, and obviously presents the species with no desiccation problems. The small size of the planktonic individuals gives them a relatively large surface for the absorption or loss of oxygen, carbon dioxide, and minerals. It also keeps them relatively immune to damage by wave action. Water may also present few nutritional problems to plants in the upper layers of the ocean, though at some times of the year certain minerals may be deficient and it should be remembered that diffusion of gases in water is considerably slower than it is through air and may provide a limiting factor to growth.

Motility may be an adaptation which enables many species at least to a small extent to find suitable habitats. It also allows for gametes to fertilise one another. Finally, through wave action and currents, the sea provides an excellent medium for the dispersal of the species.

Littoral communities

The environmental conditions of the sea shore are much more complex and varied than those of the open sea. The tides produce desiccation problems, salting-out may occur around the high water mark and wave action is much more intense. In addition the substrate is extremely varied. V. J. Chapman (1964) in a review of coastal vegetation stresses the complexity and interaction of these factors in determining the survival of a species.

The density of the sea shore vegetation on a rocky coast suggests that severe competition must have gone on between species. At the same time the zonation of species is remarkably well defined on a shore, suggesting how closely each species is adapted to its particular set of conditions. When the tide is in water provides support for the plants, the even distribution of fronds being aided by the presence of air bladders (e.g. in *Ascophyllum nodosum* and *Fucus vesiculosus*). It is not surprising therefore to find that air bladders are usually found in species commoner in the middle and upper tidal areas.

Exposure and desiccation. When the tide is out desiccation presents the same sort of problems it must have done to the first plants to colonise the land. Fig. 8.4 shows how a series of fucoid seaweeds lose water during exposure. The higher up the shore the seaweed is found, the greater its total water content, and the slower its rate of water loss. Given sufficient time, however, it will eventually lose more water than species

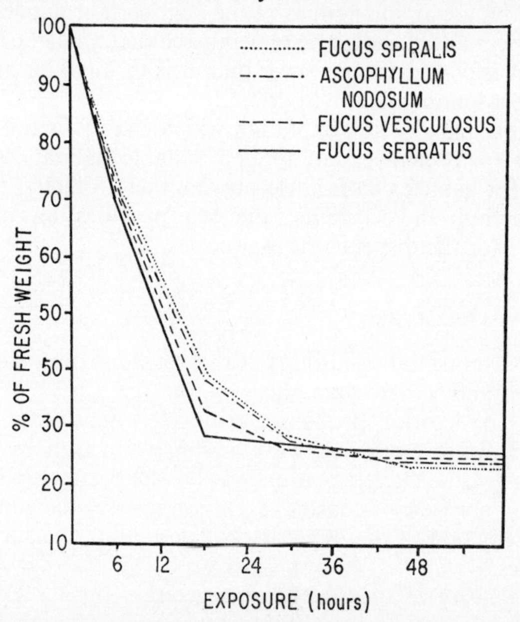

FIG. 8.4. Water loss by fucoid seaweeds during exposure

growing lower down. The rate of water loss appears to depend not only on the surface area of the plant but also on the fat content of the thallus and the thickness of the cell walls. The middle lamella of the cell wall is composed principally of the carbohydrate *mucilage* which is able to imbibe a great deal of water, and this helps to reduce water loss. Species that lose water slowly take relatively longer to reabsorb it. The reduced water availability high up the tidal zone is probably an important factor contributing to the slower growth rate of species which are found there.

Stocker and Holdheide (1937) have shown that exposure has other important effects on metabolism. Fig. 8.5 shows how exposure affects the metabolism of *Fucus serratus*. This species can continue to assimilate some time after exposure provided

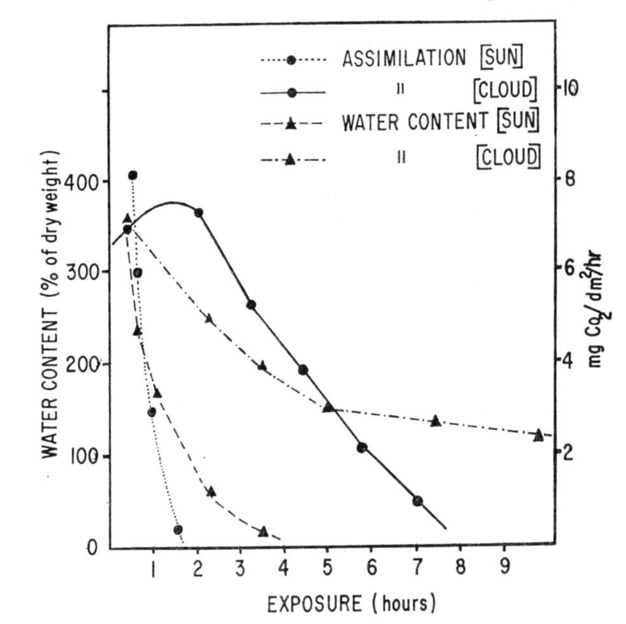

FIG. 8.5. Water loss and assimilation of *Fucus* during exposure

the conditions are cloudy and light intensity is low. Bright sunlight, presumably through increasing the temperature, increases the rate of water loss and reduces assimilation.

Salinity effects. Species found below low water mark have been found by Biebl (1938) to have a tolerance for salinity only 1·4 times greater than that of sea water. As would be expected, the nearer a species grows to the high water mark the more saline the conditions it can tolerate. *Fucus spiralis* and other species found near the high water mark can put up with a salinity three times greater than that of sea water. Green algae are also more common in brackish water than brown and red forms. It remains to be investigated whether these preferences and tolerances are osmotic or involve more complex physiology.

Wave action. Though some species (e.g. *Ascophyllum*) are unable to survive on exposed rocks and coasts, many species are extraordinarily tolerant of severe wave action. The movement of the sea also has important effects on the life cycle of the seaweeds. Normal tide movement may assist the process of fertilisation, in particular the mixing of sperms and eggs from different specimens. More severe movement will probably hinder fertilisation. Waves and currents will help to disperse zygotes, but lodgment under stormy conditions presents serious problems to the young plant, and is probably one of the most important determinants of species distribution.

Light penetration. The quality of light penetrating to different depths is another important factor determining species distribution of both open sea and littoral algal communities. The red end of the spectrum is gradually eliminated with depth. Algal pigments therefore tend to absorb light more towards the blue end of the spectrum. Fig. 8.6 shows how the distribution of radiant energy changes at lower levels; in most seas little light penetrates below thirty metres, though in extremely clear water such as the Adriatic, living algae may be found well below 100 metres.

Water and the colonisation of the land

Leaving aside for the moment algae and higher plants found in fresh water (see Chapter 9, p. 114), we can now consider the crucial phase in plant evolution of the colonisation of the land.

It is almost impossible to guess what the first true land plants looked like. Fossil evidence from the early Palaeozoic era is for the most part extremely scant. In recent surveys of the problem Sporne (1962) and Watson (1964) have simply pointed out how little is known. There is some evidence that the first land plants arose much earlier than used to be thought, probably in the Cambrian period some 500 million years ago. Marine algae certainly occurred in the period as did the first

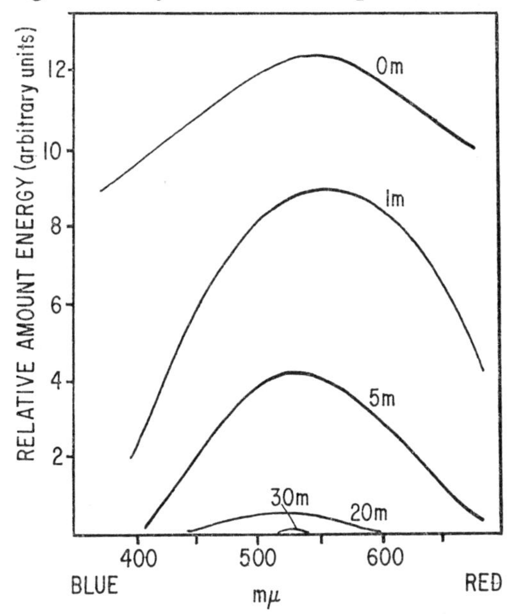

FIG. 8.6. Spectral distribution of solar energy at different depths

of the known Pteridophytes (*Aldanophyton*). Bryophyte remains are disappointly absent. Study of present day species suggests that simple thalloid liverworts such as *Metzgeria* or *Sphaerocarpus* might be similar to the earliest land plants, but the fossil record is not helpful in supporting this hypothesis. Forms similar to these, such as *Hepaticites metzgerioides* are known in the Carboniferous period, but this is many millions of years later than the first pteridophytes. Recently Neuburg

(1956) in the U.S.S.R. has described a number of mosses from the Permian period about 200 million years ago. By this time, however, the early gymnosperms were already flourishing. To suggest that the small size and poor fossil material of the bryophytes is the reason for the lack of evidence seems hardly fair as over 160 mosses and liverworts have been described from the Tertiary period.

If we accept, at least until more data is available, that some primitive pteridophytes or possibly bryophytes were among the first land plants, then the best we can do is to use modern forms to pin-point how water affects the life cycle of members of these groups.

Water and the bryophytes

Any land plant has to obtain sufficient water from the substrate to maintain its metabolism and cell turgidity. At the same time it must be able to withstand, or limit, dangerous losses through evaporation. Most mosses and liverworts live close to the substrate and are able to take up water over most of their surfaces. Rhizoids in the simpler species are probably mostly concerned with anchorage, but in some of the larger, tufted species (e.g. members of the *Polytrichaceae* such as the Australasian *Dawsonia,* which looks like a miniature fir tree a foot high) and also some of the smaller forms (*Bryum* and *Mnium*) uptake is principally through the rhizoids and transport proceeds through the primitive vascular structure of the stem to the transpiring leaves.

A number of genera possess a thin cuticle which may cut down excessive loss from the leaves and capsules. These so-called *endohydric* forms include the mosses *Bryum* and *Mnium* and the thalloid liverwort *Marchantia.* Species devoid of cuticle which absorb water extremely readily through their leaves include *Trichostomum* and *Orthotrichum* which are usually found on trees and rocks. These are *ectohydric* types. The bogmosses (*Sphagnum*) also exhibit several interesting and important features. They possess curious, non-living hyaline cells which have a great capacity for water storage. The plants also have a curious effect on the pH of the soil; through selective absorption of ions they frequently lower the pH from values around 6·0 to as low as 4·4 at the centre of the tussock.

Water is of vital importance in the bryophyte life cycle at the stage of spore germination and growth of the protonema. Perhaps the most critical phases of all concern fertilisation. The mosses and liverworts possess a wide range of sexual systems. Some are monoecious, others dioecious. Transfer of sperms within monoecious forms is obviously easier than in dioecious types but if cross fertilisation is to take place in any species the transference of sperm to the archegonium is clearly something of a problem. The process may be quite fortuitous —rain splash and small animals such as mites have been seen to transfer sperms. It seems likely that the swimming power of bryophyte sperms is poor but in some cases the antheridium releases a wetting agent with the sperms (which often stick together in a compact mass); in this way the sperms are spread much more quickly. Finally the male gametes may be attracted to the archegonium by chemical exudates, such as sugars, proteins and inorganic potassium ions which are released by many archegonia.

The next crucial phase centres on spore dispersal. In the liverworts most species depend, at least partly, on some system of *elaters*. Some (e.g. *Lophocolea*) depend on a water-rupture system. The capsules contain elaters with a bi-spiral thickening. Drying out of the capsule results in contraction of the elater through water loss. Eventually the tension in the elater is too great for the enclosed water, the water column breaks and water vapour forms; in consequence the elater untwists and expands rapidly breaking the capsule wall and releasing the spores. In many other species (e.g. *Pellia*) the elaters expand and contract due to changes in humidity and in this way dispersal is assisted. In the mosses dispersal of spores is aided by movement of the peristome teeth which normally occur in two rows around the mouth of the capsule. When the capsule is dry the peristome teeth take up the position shown in Fig. 8.7. The bent outer teeth project inwards through the gaps between the inner teeth. The up and down movements of the outer ring of teeth, induced as in *Pellia* by quite small humidity changes, help flick the spores out.

Although the bryophytes require an abundance of water at so many phases of their life cycle that 'mossy' places are usually thought of as being moist and humid, some species can thrive in remarkably dry conditions. Wall mosses, e.g.

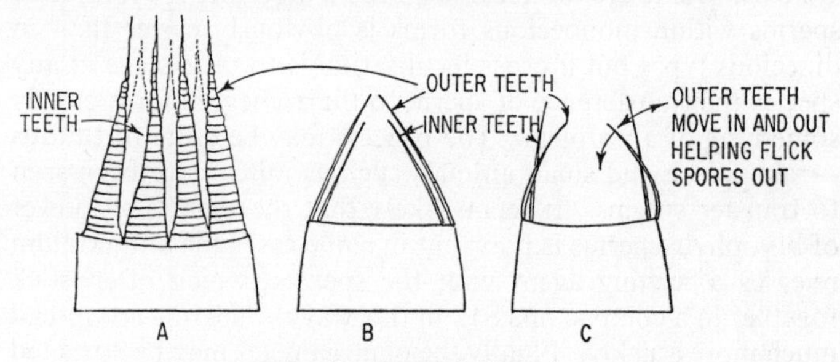

FIG. 8.7. Peristome teeth of a moss capsule

Tortula muralis can recover in a few hours or days from almost complete desiccation that may have lasted for weeks or even months. The capacity of some primitive present-day land plants to withstand such a high degree of desiccation is a truly remarkable feature.

Water and the pteridophytes

Water is as important in the life cycle of ferns as it is in mosses and liverworts, though species of *Selaginella* show features which lessen the problem of movement of the male gamete and improve the chances of the embryo developing properly. The sporophyte of the ferns has proper roots, stems and leaves, and like the higher plants, is a structure capable of efficient water economy. Thus, although the plant may be vulnerable to desiccation at some stages of its life cycle, should spore germination, prothallus formation and fertilisation be successful, the plant may be able to exist for many years in such dry situations as walls, like the wall-rue (*Asplenium ruta-muraria*) and even the more familiar male fern (*Dryopteris filix-mas*). Like many of the bryophytes, pteridophytes can also stand a remarkable degree of desiccation. Even woodland ferns are known to have existed without water for many months. Perhaps the most remarkable is the resistance of the American resurrection plant (*Selaginella lepidophylla*) which can be regarded as a xerophyte (see p. 131).

Water and the gymnosperms and angiosperms

In the bryophytes and pteridophytes the transference of the vulnerable sperms from antheridium to archegonium is clearly a most crucial step. The gymnosperms overcome this difficulty through the evolution of the waterproof pollen grain. In effect this allows for the aerial transference of male gametes, which facilitates cross fertilisation and outbreeding. The reduction of the gametophyte generation and their partial enclosure in the parent sporophyte as a protection against desiccation is another more advanced feature of the gymnosperms.

In the angiosperms the enclosing carpel wall probably protects the ovule more effectively still. The xylem of the conducting system is also more highly specialised, the tracheids of the gymnosperms being supplanted in importance in the angiosperms by xylem vessels.

Evolution in the plant kingdom has thus tended to diminish the plant's dependence on water. Today we have plant communities dominated by angiosperms and gymnosperms which often exhibit an efficient water economy. Even these terrestrial communities may also contain a wide range of primitive forms. Amongst the simplest green plants, the group of algae usually referred to as *Pleurococcus*, forms much of the green slime on trees, and algae and fungi in the form of lichens are often the primary colonisers of the driest rock habitats.

9

Water and Problems of Plant Survival

Introduction

Regardless of the taxonomic class to which they belong, plants have been grouped according to their degree of adaptation to water availability. Those living in water or very wet conditions are called *hydrophytes*, those adapted for life in equable temperate conditions of moderate humidity are called *mesophytes*, those adapted for life in drought are *xerophytes* and those which can live under salty conditions are called *halophytes*.

The hydrophytes

Lakes, ponds, and to a lesser extent rivers and streams contain floating planktonic communities similar to marine ones. The familiar euglenoids, the *Chlamydomonas-Volvox* series and the filamentous algae are often found. One obvious difference facing freshwater organisms is the problem of osmosis. The higher concentration of dissolved substances in the algal cell compared with the external solution would cause osmotic rupture of the cell unless it is able to limit its expansion. Limitation is effected by two mechanisms; the wall pressure exerted by the cell wall and the contractile vacuoles, which are capable of mechanical osmo-regulation. Many of the motile green algae (e.g. *Chlamydomonas*, *Pandorina*, *Volvox*) contain such vacuoles, but these are very often difficult to see. *Volvox* is reputed to contain five contractile vacuoles for each cell.

The converse problem is that the scarcity of some mineral

ions in freshwater may provide a limiting factor for growth. For instance the size of a population of the diatom *Asterionella* in Lake Windermere has been shown to be limited by the availability of silicon at some times of the year.

The term *hydrophyte* is often applied not only to plants living in water but also in extremely damp situations. Most of these plants possess more or less obvious adaptations to cope with the serious problems of this environment; carrying out adequate gas exchange, obtaining sufficient light for photosynthesis and depending on whether there is a current or not, on anchorage and lodgement of propagules. Many plants such as the white water-lily (*Nymphaea alba*), yellow water lily (*Nuphar lutea*) and broad-leaved pondweed (*Potamogeton natans*) have large floating leaves. Perhaps most famous of all is the giant tropical water-lily *Victoria amazonica* which has floating leaves between one and two metres across (see Plate 6). These leaves present a large surface to the light and must inevitably shade out under-water plants. They usually have their stomata confined to the upper surfaces of the leaves. Gaseous exchange and support in many of these plants is aided by the presence of numerous air spaces in the stems, leaves and even roots, which also enable them to float. This tissue is described as *aerenchyma*. Another interesting adaptation that many of these plants possess is in the length of the petioles which stretch out at an angle to the expanded leaf blade. This prevents minor changes in the level of the lake from submerging the leaf or leaving it high and dry. An interesting water plant found in tropical South America and Africa is the water hyacinth (*Eichhornia speciosa*). This possesses fleshy leaves with a curiously inflated petiole which helps to keep the leaf floating or emergent (see Plate 6). The plant increases rapidly and has become a real menace in many African lakes and waterways.

The leaves of other lake and river plants are almost always submerged. Such leaves are devoid of stomata and are usually finely dissected as in the water-milfoils (*Myriophyllum* sp.). In some species there may be alternate forms of floating leaves and submerged ones, as in some water-crowfoots (*Ranunculus aquatilis*) and in the starworts (*Callitriche* sp.). This is one of the interesting examples of the effect of the environment on the development of a plant. The survival value of dissected or long and thin underwater leaves is probably concerned with efficient light capture in the rather diffuse lighting conditions, in obtaining maximum gaseous exchange and also in helping to withstand any currents. This form of leaf is almost universal in fast-flowing streams and rivers. Establishment also presents a problem in these last habitats; many of these species spread vegetatively as well as by seeds. Small pieces of stem may become detached from the parent plant and root extremely quickly.

The halophytes

Plants adapted for living in soil conditions of high salinity are often called *halophytes*. These are found commonly in coastal areas where extensive *salt marshes* often develop. Excellent examples of these are found around the Wash where extensive mud flats have become colonised by a succession of halophytes. Less obvious examples are to be found around most estuaries and more rarely inland. Fig. 9.1 illustrates a typical zonation occurring at Keyhaven in Hampshire. Here the cord-grass (*Spartinax townsendii*) is the primary coloniser of mud and there is a great variety of plants in the drier niches. Two principal groups of factors seem to control the distribution of halophytes. These are the degree of aeration of the soil and its salinity.

Soil aeration. The ebb and flow of the tide means that the soil water saturation is always high though, perhaps surprisingly, even on a rising tide the soil water table never runs right to the soil surface, unless the area is actually inundated. The depth of this aerated layer is an important factor controlling the distribution of salt marsh plants, as many of them have their roots confined to this layer. Even though this zone is described as aerated, the amount of oxygen actually present is comparatively low and this too may be a factor influencing plant distribution. Table 9.1 below gives data of the soil atmosphere in the aerated layer surrounding the roots of the sea aster

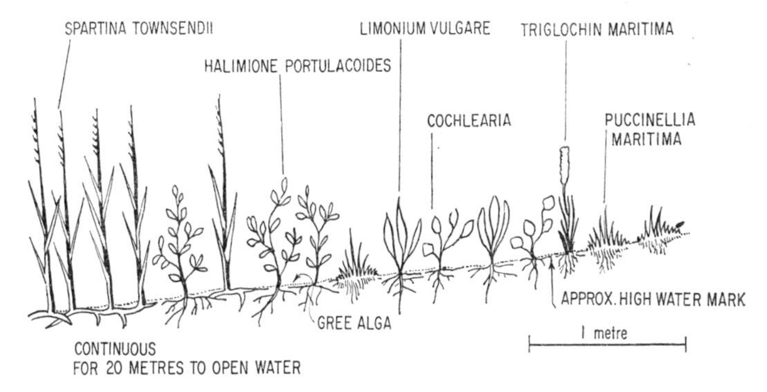

FIG. 9.1. A zonation of salt marsh plants

(*Aster tripolium*) and the matted sea lavender (*Limonium bellidifolium*), which is confined to the drier parts of salt marches and is presumably intolerant of low oxygen tensions in the soil.

TABLE 9.1

The percentage composition of salt marsh soil atmosphere

	Soil in the area of Aster tripolium	*Soil in the area of Limonium bellidifolium*
$CO_2 + H_2S$	3·3	1·2
O_2	0·9	14·0

Like many hydrophytes, halophytes also often possess aerenchyma, which may assist root growth under conditions of low oxygen concentration.

Soil salinity. In the past it was thought that the high concentration of salt and other solutes induced what has been termed physiological drought. The implication was that plants were often unable to obtain sufficient water, due to the high osmotic potential of the soil solution. It now seems that this concept must be discarded. So many salt marsh plants appear to be able to make fair use of the soil water though they may have rather high transpiration rates and seem to affect no great degree of water conservation.

The osmotic potential of the cells of many halophytes is extremely high, but it is the *salinity* of the soil, due to the presence of sodium and chloride ions, that is the important factor influencing plant survival. Chapman (1965) has emphasised the importance of these ions at various stages in the life cycle of many plants. Germination in particular is affected by salinity; on the whole the higher the salinity the lower the germination rate. The table below shows this effect in a number of plants. The tolerance of the glasswort (*Salicornia europaea*) (Plate 7 and Fig. 9.2) is high and enables it to be the primary coloniser of lower mud flats. The salinity of the soil may fall in spring-time which may provide a more

TABLE 9.2

The effect of salinity on percentage germination of salt marsh plants

	Spartinax townsendii	*Aster tripolium*	*Salicornia europaea*	*Halimione portulacoides* var. *parvifolia*
Tap water	80	45	93	25
1% NaCl	21	25	45	8·3
2% NaCl	15	10	36	0
Sea water	3	0	38	0
5% NaCl	0	0	36	0
10% NaCl	0	0	12	0

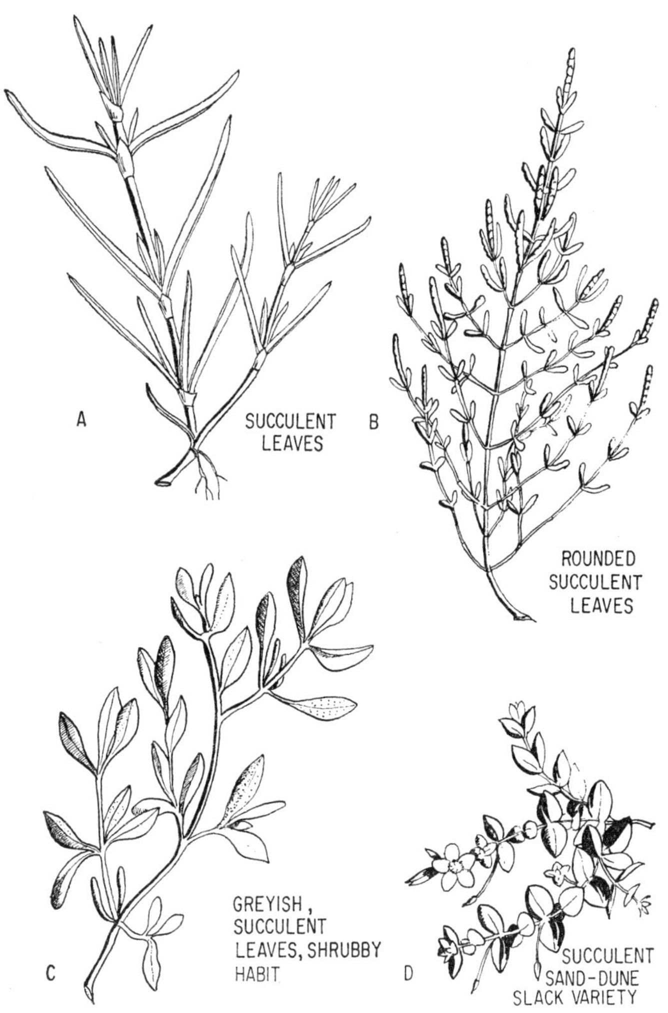

FIG. 9.2. Some salt marsh plants. A: *Spergularia marina* ($\times 1$)
B: *Salicornia europaea* ($\times \frac{1}{2}$)
C: *Halimione portulacoides* ($\times \frac{1}{2}$)
D: *Anagallis arvensis ssp. arvensis* ($\times \frac{1}{2}$)

favourable season for other species to germinate and become established.

Salinity also acts as an inhibiting factor on subsequent growth and development. *Salicornia* grows best at a salt concentration between 1·5 and 2·5 per cent while *Aster tripolium* shows its optimum growth around 0·5–1 per cent.

Plants growing in salt concentrations beyond their optima are usually stunted. There may be several causes for this. The presence of sodium ions may break down the colloidal structure of the soil. Alternatively, the plant cells may be unable to reach the abnormally high osmotic potentials which would be required to obtain water from the soil solution. Table 9.3 gives data of the osmotic potentials of a variety of plants. It must be emphasized that these figures will vary over different parts of the same plant, from plant to plant of the same species, and also at different times of the day and year.

TABLE 9.3

The osmotic potentials of various halophytes

Species	*Osmotic potential* (atm)
Glaux maritima	14·6
Triglochin maritima	24·6
Spartina patens	20·9
Salicornia europaea	39·7

Adaptations that enable halophytes to cope with problems that arise from excessively high osmotic potentials include *succulence*. This may be an environmentally induced feature as in the scarlet pimpernel (*Anagallis arvensis ssp. arvensis*) which is more common on sand dunes (see Fig. 9.2) or it may be genetically controlled as in the glassworts. It is probable that these plants compensate for their salt uptake by a corresponding water intake when the soil conditions are favourable. There seems no reason why this water should not be temporarily stored. Indeed salt marsh plants are known to have high

PLATE 7. Salt marsh vegetation near Boston, Lincs.
The early colonisers of bare mud include glasswort (*Salicornia europaea*) and cord-grass (*Spartinax townsendii*).
H. Courtesy: W. Dowdeswell

PLATE 8. Cacti in the Arizona Desert. Giant columnar cacti (*Carnegiea*) are some of the conspicuous species in the Sonora desert in Southern Arizona.
Courtesy: United States Information Service

transpiration rates so some store may be essential when the soil osmotic potential rises too high. The gradual drying up of the leaves of many succulents towards the end of the year suggests that this storage system may be essential during the hotter parts of the year. *Salicornia perennis* loses all its succulent leaves in the autumn.

Many species, e.g. *Glaux*, *Limonium*, and *Spartina*, excrete salt from special glands, though how far this prevents a serious increase in the internal concentration of salt is not known. Finally, in a number of species, the leaves simply die off when the internal salt concentration rises too high; this happens again usually towards the end of the season.

The net effect of all these adaptations is to enable halophytes to absorb and transpire much the same quantity of water per unit area plant as an ordinary mesophyte.

The xerophytes

Plants adapted to conditions of drought or water scarcity are called *xerophytes*. This term includes a wide range of plants, some with little obvious adaptation, others highly modified. Plants which resemble xerophytes but which are intolerant of drought are called *Xeromorphs*. Water shortage occurs intermittently in almost any part of the world and mesophytes too must have some form of drought protection. The most fascinating xerophytes are those found in or near deserts. No plant is capable of surviving indefinite drought, it is only where water is present either deep in the soil or occasionally from rain that some form of plant life can exist.

Desert vegetation of the Sahara and Gobi is usually composed of plants with rather dry, hard leaves; succulents are relatively rare. This is in sharp contrast to the deserts of America and South Africa where succulent species are much more common. The true cacti of the *Cactaceae* were formerly confined to the New World but are now found commonly in many parts of the world. In Australia the encroachments of the prickly pear (*Opuntia inermis* and *Opuntia stricta* in particular) on agricultural land have proved to be an extremely serious problem. In 1925 an area of some 60 million acres—an area about six times that of Switzerland—had become infested by prickly pears. At this time biological control using the large Argentinian moth, *Cactoblastis cactorum* began to be used. The larvae burrow into the fleshy stem segments and so effectively control the weed. The succulents of Southern Africa, particularly the genus *Euphorbia*, sometimes look exceedingly like the cacti and provide an interesting example of convergence in evolution.

It is possible to classify the xerophytes on the basis of their anatomical and physiological characteristics, but perhaps the simplest course is to separate from the start the group of xerophytes known as *drought evaders* which are killed by drought and survive such times as seeds or spores. The remaining group, adapted to drought survival can be called the '*drought endurers*'.

The drought evaders. Annuals and ephemerals form much of the vegetation that springs up almost overnight after a heavy desert shower. They germinate, grow, flower and set seed in only four to six weeks. They resemble mesophytes in many ways and are adapted for life in deserts simply by tiding over the drought period in the form of a seed (or more rarely a spore). Many familiar cultivated annuals come into this class; the Californian poppy (*Eschscholtzia californica*) is a well-known example.

In Britain a number of small annual plants such as the whitlow grass (*Erophila verna*) which is common on walls and paths have a rather similar life-cycle. Seeds germinate in the late autumn and growth takes place during the damper and cooler months of winter and spring. Flowering is usually over by May and seeds are set before the drier summer begins. It is interesting to note that the dormancy of the seed seems to be to some extent under internal control. Laboratory germination experiments have shown that seeds do not germinate readily even if given suitable conditions in May soon after they are set.

Maintenance of water uptake by drought endurers. Any plant that is taking up water freely could hardly be said to be living under conditions of acute water scarcity. Yet there are a number of desert plants with exceedingly deep roots which enable them to tap the water table as and when it is available. Many Mediterranean trees and shrubs such as the acacia and oleander come into this class. One of the problems here clearly relates to establishment; it may be many years before the plant's roots are able to grow down to the water. Other desert plants, particularly the cacti, have shallow spreading root systems which enable the plant to obtain maximum water when rainfall occurs. However, a surface root system of this sort is bound to become almost completely desiccated in dry weather.

As would be expected, most, but by no means all, xerophytes have high osmotic potentials and so can extract water more readily from drying soils. Shrubby species often have osmotic potentials in excess of 25 atmospheres, but many succulents have quite low values.

Storage of water. Many xerophytes have rounded, fleshy leaves and stems in which considerable quantities of water may be stored. These structures are composed mainly of large parenchymatous cells containing mucilage. This probably helps the cells imbibe water. The ice plant (*Mesembryanthemum crystallinum*) possesses large bladder-like epidermal cells in which water is stored.

Reduction of water loss. There are many adaptations which help prevent excessive water loss. One of the most obvious involves a reduction in the surface area of the plant. If this goes along with an increase in volume the plant will also become more succulent. Leaves, stems, or both may be modified in this way, in consequence drought endurers have often been classified into stem and leaf succulent series.

Reduction in surface area; the stem and leaf succulent series. In the leaf succulent series (see Fig. 9.3), perhaps the least well-adapted structurally are species of stonecrop (*Sedum*) and *Bryophyllum*. Here, although the leaves are swollen, there is probably little reduction in transpiring surface. Species of

Crassula, *Mesembryanthemum* and Hottentot fig (*Carpobrotus*) are better adapted; the internodes are shorter and the leaves more succulent and tightly packed. Many succulents have a

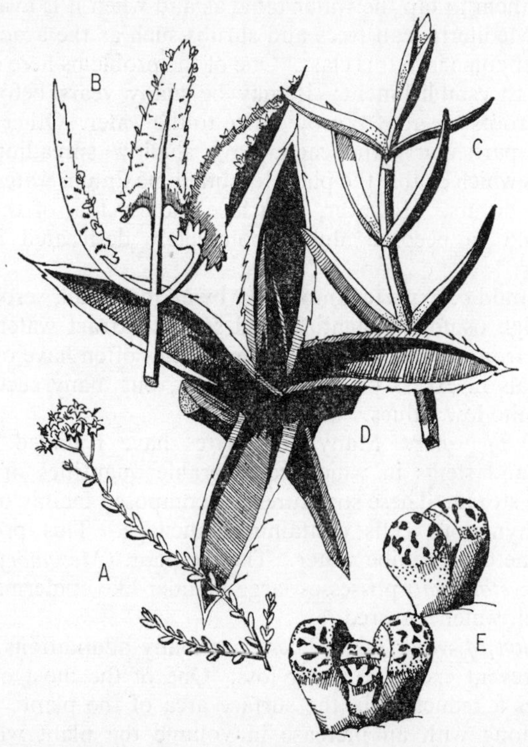

FIG. 9.3. A series of leaf succulents. A: *Sedum album* B: *Bryophyllum daigremontianum* C: *Carpobrotus edulis* D: *Agave americana* E: *Lithops helmuti* (approx. × ½)

rosette form with relatively few large succulent leaves. Members of the lily family, e.g. *Aloe*, *Haworthia* and the century plant (*Agave americana*) in the *Amaryllidaceae* are in this group. Many of these plants are often a feature of gardens and hillsides at the south of France. Some of the most fascinating leaf succulents are undoubtedly in the genus *Lithops*, like *Mesembryanthemum*, a member of the *Aizoaceae*. These so-called pebble plants are natives of the Kalahari desert in South Africa where they resemble stones so closely that they have sometimes been called 'mimicry plants'. The stem is extremely short and the leaves very swollen, the whole being sunk in the ground so that the tops of the leaves seldom rise much above ground level. The upper surface of the leaf is often more or less transparent and allows light to reach the photosynthesising cells which line the leaf edges. The growing point is also well-protected by the leaves and is in any case often two centimetres below ground. In these species the transpiring surface is clearly minimal.

Succulents are an obvious source of food and water for desert animals. Some like *Lithops* may be well camouflaged but they are known to be the food of baboons and bustards. Perhaps the stories of ostriches with their heads buried in the sand have their source here! Other species like *Agave* are well protected with spines.

In the stem succulent series (see Fig. 9.4) the least modified are perhaps the candle plant (*Kleinia articulata*) and the Barbados gooseberry (*Pereskia*). These have both true leaves and moderately swollen stems. The candle plant loses its leaves in times of drought but continues photosynthesis through the stem. The Barbados gooseberry is a curious member of the *Cactaceae*, for it is a shrub, but it possesses the typical woolly areole and spines in the leaf axil. Some of the most famous stem succulents are similar to *Pereskia* without the expanded leaf. In the columnar cacti the stem is only rarely branched; the giant species of *Carnegiea* and *Cereus* are perhaps some of the best adapted xerophytes known. Other forms such as *Echinocactus* are almost spherical in form and like most of these species are well protected with spines. Some of these resemble giant pin-cushions sometimes half a metre across.

In some stem succulents there are grooves and ridges which may permit efficient gas exchange and photosynthesis without

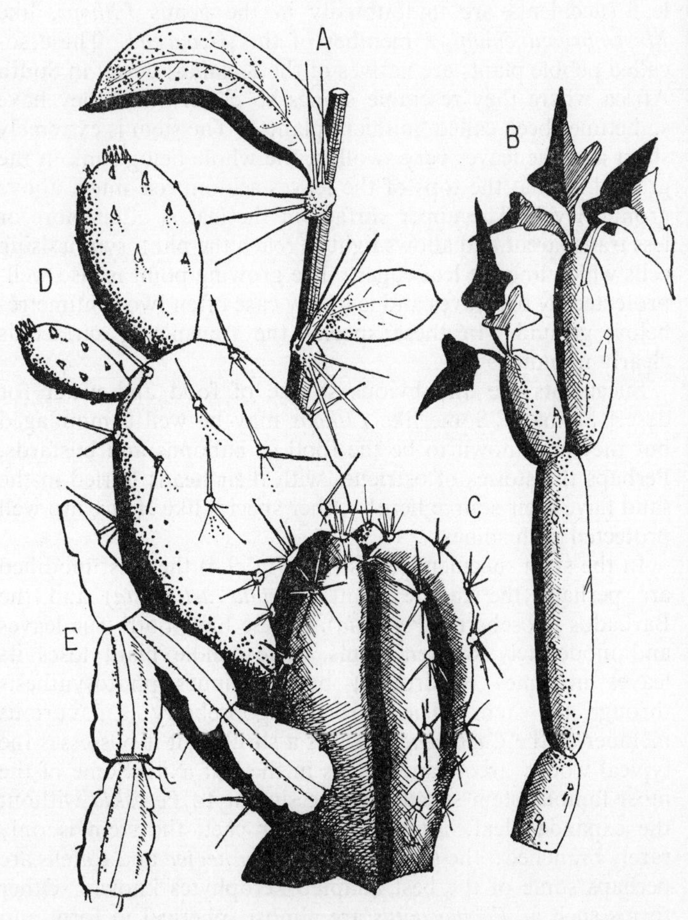

FIG. 9.4. A series of stem succulents. A: *Pereskia bleo* B: *Kleinia articulata* C: *Cereus* D: *Opuntia* E: *Zygocactus truncatus* (approx. × $\frac{1}{2}$)

disrupting the plant's water economy. In other cases though, the stem segments are flattened and leaf-like. Species of prickly pear (*Opuntia*) form large bushy plants with rounded leaf segments up to 40 cm long. The cacti also include epiphytic and hanging species (*Zygocactus* and *Epiphyllum*) which, although they superficially resemble some xerophytes, are not particularly well adapted for drought resistance and live in quite different habitats such as tropical forests.

Apart from these succulents a number of other adaptations help in the lowering of the transpiring surface. Some of the most effective are underground storage organs. Many bulbous plants have their centre of distribution in countries with a Mediterranean type of climate—areas where the summers are exceedingly hot and dry. At the end of spring most species die down to their underground bulbs and corms which hold considerable reserves of water in their fleshy stems and leaves. These are protected by scale leaves and so the plant is able to withstand complete drought for many months.

Many semi-desert areas, particularly in steppe country, are sparsely covered by plants of low-growing cushion or trailing habit which have a relatively low transpiring surface. *Artemisia frigida* which occurs both in North America and Siberia is a low woolly shrub which covers many thousands of acres of steppe country. The American sage brush (*Artemisia tridentata*) although quite a large shrub has a similar structure and a particularly well developed root system. Fig. 9.5 shows a transect through a hillside in southern Spain; the abundance of bulbous plants and small shrubs is particularly noteworthy.

Quite a number of xerophytes shed their leaves in times of drought. Usually the older leaves are shed first and the most vital ones near the apex are the last to go. The South African elephant's foot *Dioscorea elephantipes* combines both adaptations as it has a large tuber and loses much of its three-metre leafy stem in times of drought. In some plants the survival of younger leaves may depend on the withdrawal of water from fruits and older leaves.

Modifications to the epidermis. Apart from reduction in the surface area of the plant, transpiration can also be lessened by modifications of the stem and leaf surface. The deposition of cutin, a fatty derivative, gives many plants a shiny surface; this is particularly clearly seen in *Crassula* and *Opuntia*. Sometimes the cuticle may have a waxy layer which gives the plant a bluish bloom (e.g. in *Kleinia* and *Cereus*). These substances probably reduce cuticular transpiration. Many xerophytes, e.g. *Stachys kotschyi* and *Artemisia frigida* appear

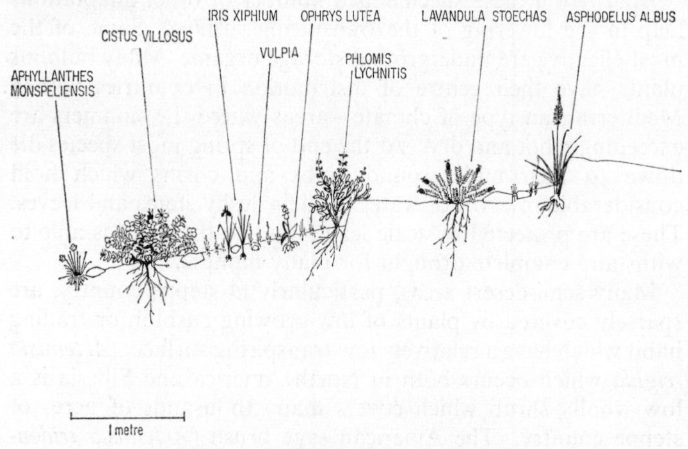

FIG. 9.5. A Mediterranean hillside

white due to the presence of white hairs over the leaf surface. At first sight these might appear to provide a greater surface area for transpiration and indeed many of these xerophytes have quite high transpiration rates (see Table 9.4). On the other hand it could be argued that by creating a still air layer around the leaf, the hairy covering may reduce transpiration.

Transpiration is probably more strongly limited by the number, size and arrangement of stomata. Compared with mesophytes the number of stomata in the xerophytes, though not necessarily the size, is usually reduced. Many of the former group have between 400 and 1000 stomata per sq. mm. Cacti such as *Opuntia ficus-indica* have only about 100 per sq. mm and many species have even less.

Quite often the stomata may be sunk in pits as in the pine trees and *Dasylirion*. In *Dasylirion* there are two cavities outside the guard cells (see Fig. 9.6) which must effectively

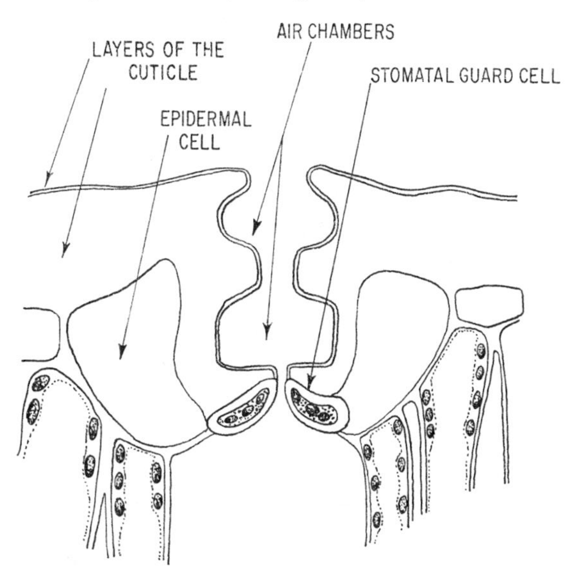

FIG. 9.6. Sunken stoma of *Dasylirion* (highly magnified)

reduce the rate at which water molecules diffuse through the stomatal system. Similarly water loss may be reduced by hairs immediately around the stoma. The marram grass (*Ammophila arenaria*) has hairs around the sunken stomata and in addition curls up its leaf round the stomata in times of drought (see Fig. 9.7).

A number of xerophytes, particularly members of the *Aizoaceae* such as *Mesembryanthemum* have an *inverted stomatal rhythm*, the stomata being closed by day and open at night. As most of the water lost from a plant is through the stomata, this physiological adaptation must be of considerable importance, provided that sufficient carbon dioxide can be taken up for photosynthesis.

Another physiological point that is certainly of interest and may be important is the fact that so many xerophytes are aromatic. Thymes, lavenders, and various labiates are characteristic of Mediterranean vegetation. It has been suggested,

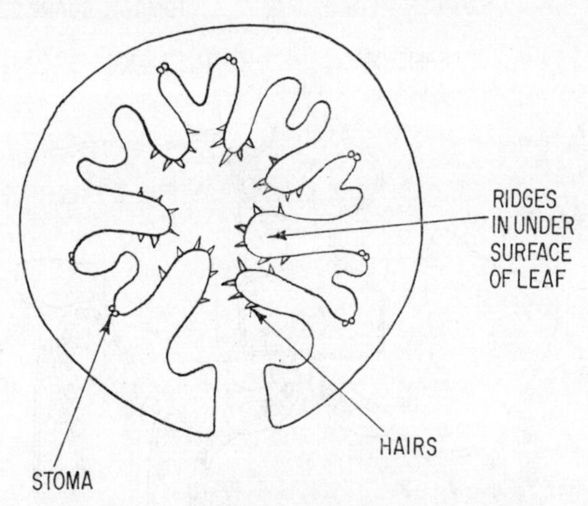

FIG. 9.7. Transverse section of the dry leaf of marram grass (highly magnified)

though not proved, that the production of aromatic volatile oils in some way screens the plant and helps prevent excessive transpiration.

Some examples of transpiration rates of xerophytes and mesophytes are given in Table 9.4.

This shows that many woolly-leaved and succulent species in particular have fairly high transpiration rates, but xerophytic plants showing very evident adaptations to drought resistance do in fact have lower transpiration rates than the less obviously adapted species.

While this is particularly marked in species with sunken stomata and in the cacti, MacDougal and Spalding (1910) have

TABLE 9.4

The intensity of transpiration and the rate of expenditure of the water store of a number of xerophytes and mesophytes

Species	*Intensity of transpiration* (g/dm²/hr)	*Rapidity of expenditure of water store* (Loss/hr as % total water content)	*Character of the leaves*
XEROPHYTES			
Sedum maximum	2·8	8	Succulent
Gypsophila acutifolia	5·4	20	Hard, fleshy
Falcaria vulgaris	13·7	87	Hard, covered with wax
Stachys kotschyi	12·7	119	Densely hairy
MESOPHYTES			
Campanula rapunculoides	4·8	36	Herbs with slightly hairy leaves (shade plants)
Viola odorata	4·0	58	Herbs with slightly hairy leaves (shade plants)
Erodium ciconium	9·2	83	Quite hairy leaves (sun plant)

demonstrated how slowly some cacti lose water. A large specimen of *Carnegiea* (Plate 8) weighing 45 kg lost only 23 per cent of its weight when left unwatered for a year and can apparently lose 63 per cent of its water content without serious injury. Similarly a large *Echinocactus* kept without water was weighed each year for six years and lost water extremely slowly after the first three years (see Fig. 9.8).

Desiccation. The ability of xerophytes to 'come to life' so quickly and to survive in a high state of desiccation is as remarkable as the viability of dormant seeds. While it is probably true that the protoplasm of many of the xerophytes already described can stand varying degrees of desiccation, the ability of some species to recover after severe dehydration is little short of spectacular. The creosote bush (*Covillea glutinosa*) of the North American desert (see Fig. 9.9) has small leathery leaves covered with a sticky resin. When drought occurs, these wilt, shrivel, and become brownish. A few of the lower leaves may be shed. Nevertheless most of the leaves remain alive, even though the plant may be obtaining no water from the soil. Recovery is extremely rapid and assimilation can recommence within a few

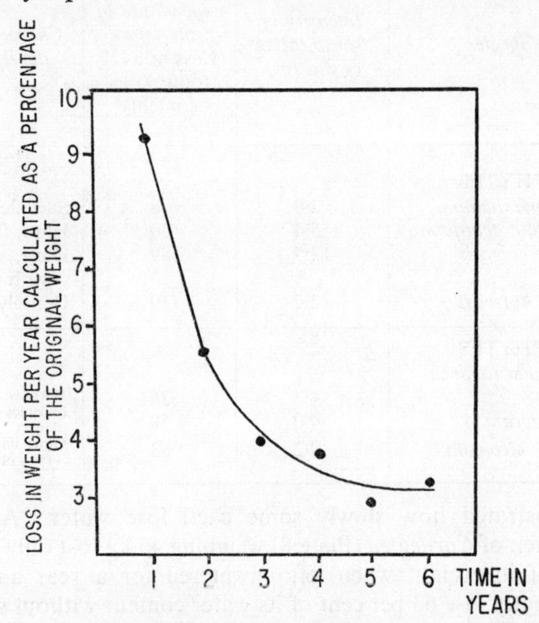

FIG. 9.8. Water loss from *Echinocactus*

days. Yapp (1929) gives details of an equally remarkable plant, *Myrothammus flabellifolia*, which lives in crevices of granite rocks in the Matopo hills in Rhodesia. After four months of drought these plants were dry and brittle, their water content being less than 10 per cent of normal. Yapp kept twigs of this plant in an air-dry environment for a further thirteen months. They were then put in water and became green and expanded overnight. However, he did not investigate the metabolic activity of these recovered shoots and by analogy with other plants such as the bryophytes the plants may appear healthy though in fact may be dead. Another important point is that the leaves may be more tolerant of

FIG. 9.9. Shoot of the creosote bush (*Covillea glutinosa*) (× $\frac{2}{3}$)

desiccation than other parts of the plant such as the root hairs; death of the latter may cause failure of the whole plant.

Some aspects of the water relations of lower plants have already been discussed (see p. 110) and there is no doubt that many mosses, liverworts, and lichens can stand high degrees of desiccation for long periods. Bristol (1916) showed that resting but viable moss protonemata could be obtained from air-dry soils that had been kept in stoppered bottles for nearly 50 years! In most higher plants it is usually only the seed that is capable of withstanding really prolonged desiccation.

Yapp (1916) also showed that relatively small, less specialised, cells are much more tolerant of dry conditions. When leaves are exposed to hot, dry winds areas of parenchymatous

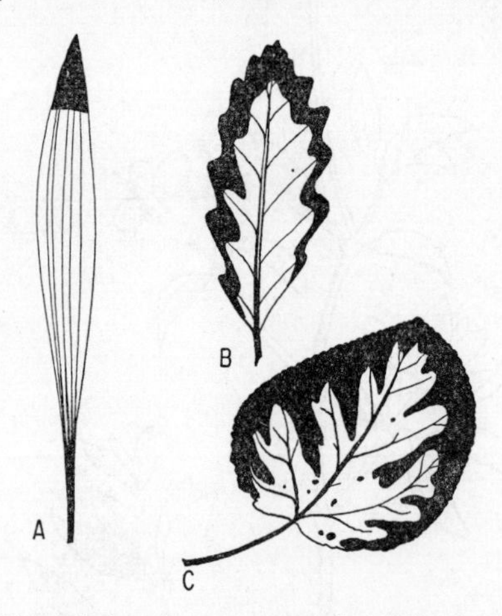

FIG. 9.10. Leaf desiccation patterns. A: Plaintain B: Oak tree C: Poplar

and thin-walled tissue are most seriously affected. These areas were usually on the leaf margins, near the leaf apex or between the veins (see Fig. 9.10).

Cell structure in relation to desiccation and frost damage

Gardeners 'harden off' plants to increase their frost tolerance. In the same way plants may acquire some degree of drought tolerance, if the water supply is kept fairly low. To some extent, the two phenomena go hand in hand as the better hardened-off a plant the more tolerant it may be of both drought and frost.

In both instances the un-hardy plant probably succumbs due to damage to the larger, highly vacuolated cells. In the case of drought, water loss may result in cellular distortion and protoplasmic rupture, even before the vacuole has completely evaporated. Low temperature may cause ice crystals to form which may also kill the cell. Tolerance is often improved as the concentration of dissolved substances in the cell is increased; gradual decrease in temperature and water supply reduce the size of the vacuole and favour the accumulation of cell colloids such as starch, fats, and proteins in both cytoplasm and vacuole. This greatly lessens the likelihood of cell disorganisation. Iljin (1912) demonstrated this effect. When cells of *Iris* mesophyll were soaked in sugar solutions of increasing strengths their tolerance of dry conditions could be increased. Maximov also demonstrated that the frost tolerance of the cells could be improved at the same time.

TABLE 9.5

The effect of glucose on the humidity tolerance of Iris *mesophyll cells*

Pre-treatment of the cells	*Relative humidity* (per cent)	*Result*
Washed in water	99	Dead
Soaked in 0·1 M glucose	97	Alive
Soaked in 0·2 M glucose	93	Alive
Soaked in 0·5 M glucose	90	Alive

Horticultural selection of plant varieties for drought and frost resistance has of course gone on for hundreds of years. Different strains of the same species can exhibit markedly different tolerances; the greater frost tolerance of red cabbage compared with the ordinary green varieties is well known.

The work of Tumanov (1927) on the drought resistance of different varieties of spring wheat all kept in a state of drought for two weeks show the form such work has taken. Differences in survival rate seem to relate for the most part to internal conditions occurring in the plants' cells. The table below summarises some of Tumanov's results.

TABLE 9.6

Drought resistance in varieties of spring wheat

Name of variety	*Number of pure line*	*The number of plants*		*Percentage of survivals*
		Total	*Survived*	
Ferrugineum rossicum	Tulun 120/32	33	31	94
Caesium	Omsk 111	11	9	82
Lutescens	Marquis	35	27	77
Ferrugineum rossicum	Tulun 916/4	18	9	50
Ferrugineum rossicum	Tulun 324	19	8	42
Anglicum	Pusa 4	34	8	23

Here there is clearly an interesting tie-up between genetic and physiological work which suggests several profitable future lines of research.

Bibliography

BARON, W. M. M. *Organisation in Plants* (Arnold, London, 1963).

BOLLARD, E. G. 'Transport in the Xylem,' *Ann. Rev. Plant Physiology*, 1960, Vol. 11, p. 141.

BRIGGS, L. G. and SHANTZ, H. L. 'The Water Requirements of Plants as influenced by Environment,' *Proc. 2nd Pan-American Scientific Congress*, 1917.

BROWN, H. T. and ESCOMBE, F. 'Static Diffusion of Gases and Liquids in relation to the assimilation of Carbon Dioxide and Translocation in Plants,' *Phil. Trans. Roy. Soc.*, 1900, **B.193.**

BUSWELL, A. M. and RODEBUSH, W. H. 'Water,' *Scientific American*, April, 1956.

CHAPMAN, V. J. *Coastal Vegetation* (Pergamon Press, Oxford, 1964).

CRAFTS, A. S., CURRIER, H. B. and STOCKING, C. R. 'Water in the Physiology of Plants,' *Chronica Botanica*, 1949 (Waltham, Mass.).

DARWIN, F. 'On the Relation between Transpiration and Stomatal Aperture,' *Phil. Trans. Roy. Soc., Lond.*, 1916, **2–70**, 413.

DIXON, H. H. *Transpiration and the Ascent of Sap in Plants* (Macmillan, London, 1914).

ESAU, K. *Plant Anatomy* (Wiley, New York, 1953).

ESAU, K., CURRIER, H. B. and CHEADLE, V. I. 'Physiology of Phloem,' *Annual Review of Plant Physiology*, 1957.

GREENIDGE, K. N. H. 'Ascent of Sap,' *Ann. Rev. Plant Physiology*, 1957, Vol. 11, p. 237.

JENSEN, W. A. and KAVALJIAN, L. G. *Plant Biology Today* (Macmillan, London, 1965).

JUNIPER, B. E. 'The Surfaces of Plants,' *Endeavor*, 1959, XVIII, No. 69.

KAMIYA, N. 'Physics and Chemistry of Protoplasmic Streaming,' (*Annual Review of Plant Physiology*, 1960).

LEYTON, L. and JUNIPER, B. E. 'Cuticle Structure and Water Relations of Pine Needles,' *Nature*, London, 1963.

MAKERETH, F. J. H. *Water Analysis for Limnologists* (Freshwater Biological Association, Ambleside, 1957).

MARTIN, E. V. *Studies of evaporation and transpiration under controlled conditions* (Carnegie Inst. Wash. Publ. 550. Washington, 1953).

MAXIMOV, N. A. (Ed. R. H. YAPP). *The Plant in Relation to Water* (Allen and Unwin, London, 1929).

MEYER, B. S. and ANDERSON, D. B. *Plant Physiology* (Van Nostrand, New York, 1952).

NOEL, A. R. A. 'Some new techniques in plant physiology,' *School Science Review*, No. 142, 1959.

PAUL, JOHN. *Cell Biology* (Heinemann Educational Books, London, 1965).

RUSSEL, E. J. *Soil Conditions and Plant Growth* (Longmans, London, 1953).

RUTTER, A. J. and WHITEHEAD, F. H. 'The Water Relations of Plants,' *British Ecological Society. Symposium No. III* (Blackwell, Oxford, 1963).

SOCIETY OF EXPERIMENTAL BIOLOGY. *S.E.B. Symposium Number XIX. The State and Movement of Water in Living Things* (Cambridge University Press, Cambridge, 1965).

STREET, H. E. *Plant Metabolism* (Pergamon, Oxford, 1963).

WEATHERLEY, P. E. 'Some theoretical considerations of Cell Water Relations,' *Annals of Botany N.S.*, 1952, Vol. XVI, No. 62, p. 41.

ZIMMERMAN, M. H., 'Transport in the Ploem,' *Ann. Rev. Plant Physiology*, 1960, Vol. 11, p. 167.

ZIMMERMAN, M. H. 'How Sap Moves in Trees,' *Scientific American*, March, 1963.

Acknowledgements

Figs. 1.1, 1.2 redrawn after Buswell, A. M. and Rodebush, W. H., 'Water,' *Scientific American*, April 1956.

Fig. 3.3 redrawn after Street, H. E., *Plant Metabolism*, Pergamon, Oxford, 1963.

Figs. 2.3, 4.1, 4.2, 9.6, 9.10, Tables 2.3, 2.4, 4.1, 5.1, 5.2, 9.4, 9.5, 9.6 after Maximov, N. A., *The Plant in Relation to Water*, Allen and Unwin, London, 1929, p. 56.

Table 2.1 from Leyton, L. and Juniper, B. E., 'Cuticle Structure and Water Relations of Pine Needles,' *Nature*, **198,** 4882, 1963, p. 770.

Table 2.2 from Monteith, J. L., 'Dew: Facts and Fallacies' in *The Water Relations of Plants*, edited by Rutter, A. J. and Whitehead, F. H., Blackwell, Oxford, 1963, p. 54.

Fig. 3.2 redrawn after Weatherley, P. E., 'Some Theoretical Considerations of Cell Water Relations,' *Annals of Botany N.S.* Vol. XVI, No. 62, April 1952, p. 141.

Fig. 3.5 redrawn after Skoog, F. *et al.* 'Effects of Auxin on rates, periodicity and osmotic relations in exudation,' *Amer. J. Bot.* **25,** 752, 1938.

Figs. 4.6, 4.11, 5.9, Table 5.4 redrawn after Willis, A. J. and Jefferies, R. L., 'Investigations on the water relations of sand-dune plants under natural conditions,' *The Water Relations of Plants*, Rutter, A. J. and Whitehead, F. H., 1963, p. 172.

Fig. 4.7 redrawn after data of Martin, E. V., 'Studies of evaporation and transpiration under controlled conditions,' *Carnegie Inst. Wash. Publ.* 550 (Washington, 1943).

Figs. 4.9, 4.10, 6.3, Tables 3.1, 6.1 redrawn after Meyer, B. S. and Anderson, D. B., *Plant Physiology*, Van Nostrand, New York, 1952.

Figs. 5.2, 5.3, Table 5.3 adapted from data of Brown, H. T. and Escombe, F., 'Static diffusion of gases and liquids in relation to the assimilation of carbon and translocation in plants,' *Phil. Trans. Roy. Soc.*, London, **193B,** 223–291.

Fig. 5.5 (grass-type) redrawn after Essau, K., *Plant Anatomy*, Wiley, New York, 1953, p. 149.

Fig. 5.7 redrawn after Noel, 'Some new techniques in Plant Physiology,' *School Science Review*, No. 142, 1959, p. 497.

Fig. 5.8, 'A simple potometer for measuring the resistance to air flow offered by stomata of green leaves,' redrawn after Meidner, H., *School Science Review*, No. 161, 1965, p. 149.

Figs. 5.10, 5.11, Table 5.4, after Geyer, T. A., 'Investigations into the behaviour of stomata in the field,' *School Science Review*, No. 153, 1963, p. 390.

Figs. 6.7, 6.8, 6.9, Tables 6.2, 7.1, data of Salisbury, F. B., 'Translocation: the movement of dissolved substances in plants,' from Jensen, W. A. and Kavaljiah, L. G., *Plant Biology Today*, Macmillan, London, 1963, p. 87.

Figs. 7.2, 7.3, 7.4, 7.5, 7.6, 7.7, 7.8, 7.9, 7.10. 7.11, 7.12, 7.15 redrawn after Weatherley, P. E., 'The pathway of water movement across the root cortex and leaf mesophyll of transpiring plants,' in Rutter, A. J. and Whitehead, F. H. as 4.6 above.

Figs. 7.13, 7.14 redrawn after Zimmermann, M. H., 'How sap moves in trees,' *Scientific American*, March 1963, p. 3.

Figs. 8.1, 8.2 redrawn after the BSCS Blue version, *Molecules to man*, Arnold, London, 1963, p. 95.

Fig. 8.3 redrawn after Stanier *et al.*, *General Microbiology*, Macmillan, London, 1963, p. 415.

Figs. 8.4, 8.5, 8.6, Tables 9.1, 9.2, 9.3 redrawn after Chapman, V. J., *Coastal Vegetation*, Pergamon, Oxford, 1964, p. 59.

Fig. 9.9 redrawn after James, W. O., *Plant Physiology*, 5th edn., Oxford, 1957, p. 173.

Index